POLY(TETRAHYDROFURAN)

Polymer Monographs

Edited by M. B. Huglin, University of Salford

A series of short monographs, each dealing with one specific polymer. They are written by experts from leading research laboratories, and cover equally basic scientific information on a polymer and information pertinent to its practical utilization.

International Standard Serial Number (ISSN): 0275-5777.

Volume 1
POLY(1-BUTENE)—Its preparation and properties
I. D. RUBIN, *Texaco Inc.*

Volume 2
POLYPROPYLENE
H.P. FRANK, *Osterreichische Stickstoffwerke*

Volume 3
POLY(VINYL CHLORIDE)
J. V. KOLESKE and L. H. WARTMAN, *Union Carbide Corporation*

Volume 4
POLY(VINYL ALCOHOL)—Basic properties and uses
J. G. PRITCHARD, *West Ham College of Technology*

Volume 5
POLYVINYLIDENE CHLORIDE
R. A. WESSLING, *Dow Chemical Co.*

Volume 6
POLY(N-VINYLCARBAZOLE)
J. M. PEARSON, *Eastman Kodak Co.* and M. STOLKA, *Xerox Corporation*

Volume 7
HEPARIN (AND RELATED POLYSACCHARIDES)—Structural and functional properties
WAYNE D. COMPER, *Monash University*

Volume 8
POLY(TETRAHYDROFURAN)
P. DREYFUSS, *The University of Akron*

POLY(TETRAHYDROFURAN)

P. DREYFUSS
Institute of Polymer Science
The University of Akron

GORDON AND BREACH SCIENCE PUBLISHERS
New York London Paris

Gordon and Breach, Science Publishers, Inc.
One Park Avenue
New York, NY 10016

Gordon and Breach Science Publishers Ltd.
42 William IV Street
London, WC2N 4DE

Gordon & Breach
58 rue Lhomond
75005 Paris

Library of Congress Cataloging in Publication Data

Dreyfuss, P. (Patricia),
Poly(tetrahydrofuran)

(Polymer monographs; v. 8)
Includes bibliographical references and indexes.
1. Tetrahydrofuran. 2. Furan resins. I. Title.
II. Series.
TP1180.F8D73 668.4'2 81-4273
ISBN 0-667-03330-3 AACR2

Library of Congress catalog number 81-4273. ISBN 0 677 03330 3. ISSN 0275-5777. Printed in the United States of America.

Contents

Series Editor's Preface

"*If a book is worth reading, it is worth buying.*" (John Ruskin)

Two trends, which are not especially favourable to a series such as this have become noticeable in recent years, *viz*—first, the reluctance of the polymer industry to invest in the development of entirely new homopolymers and, second, the increasing costs of scientific book production. Enquiry and feedback revealed, perhaps surprisingly, but nonetheless emphatically, a genuine need for authoritative up-to-date treatments on several existing polymers. Consequently, each volume is devoted to one specified polymer. The interests of an industry or an institution usually dictate that research and/or development be conducted over an extended period of time on the particular polymer of relevance. It is hoped that individuals will be able to select their specific volume from Polymer Monographs and thus be freed from the inconvenient and unfair obligation to pick sections out of the commoner, large treatises devoted to classes of polymer.

With regard to the topics themselves, proven useful application is considered a *sine qua non*, and hence macromolecules of purely academic interest are excluded from this series. An innovation is the inclusion of biopolymers and their synthetic prototypes. Since polymer science has now become truly interdisciplinary in scope, the monographs are addressed to a broad spectrum of potential readers.

These volumes are short, the aim being to present the maximum of current information and in the minimum of space. Fortunately the authors, who are all pre-eminent in their respective fields, have not only complied with this difficult and stringent condition, but have succeeded in doing so without sacrificing readability. The publishers have endeavoured to make these books available at a price which is reasonable by present standards. They and I trust these volumes will prove useful, and welcome comments and suggestions for future topics.

MALCOLM B. HUGLIN

Author's Preface

The polymerization of tetrahydrofuran can be beautifully simple and often serves as a model for the study of oxonium-ion polymerization and for the examination of the accuracy of various theoretical predictions. It can also be frustratingly complex. The many factors that affect the polymerization, the unexpectedly large effects that result from simple changes such as in solvent or temperature, and the properties of the polymers and copolymers formed constitute the subject of this monograph.

Every reviewer writes with a particular audience in mind and selects material to be included in a given work in terms of that audience. My monograph is directed towards both novices and experts in the field of poly(tetrahydrofuran). When choices had to be made to keep the length within reasonable limits, I elected to emphasize recent publications at the expense of older, previously reviewed, literature. I have tried to be comprehensive in scope but not all-inclusive in material.

The monograph begins with an introduction to the polymer and its monomer, tetrahydrofuran. It leads the reader through the labyrinth of synthetic processes now available. It discusses kinetics of polymerization, properties of the polymer and structure–property relationships, and devotes two whole chapters to block and graft copolymers containing poly(tetrahydrofuran); one of these deals with the commercially important polyurethanes and polyesters. At every stage I have superimposed a critical and personal evaluation of existing data on an exposition of the current state of the art. If this monograph helps novices appreciate the many factors that need to be considered to carry out a well-controlled tetrahydrofuran polymerization that will give the kind of polymer they desire, and if present experts disagree with some of my conclusions enough to run new, inspired experiments, my efforts will have been rewarded.

I am indebted to many people for help in bringing this monograph to completion. First and foremost, I wish to acknowledge the help of my colleague and spouse, M.P. Dreyfuss, who has kept an ever-vigilant eye out for information that might be useful to me and who has read, critiqued and proofread most of the chapters. I am grateful to Marilyn Israel for typing the manuscript, to Perry Marteny for preparing many of the drawings and to Larry J. Wilson for help in preparing the final figures. I thank C. S. Schollenberger for supplying much information for the polyurethane section, M. B. Huglin for providing many papers related to solution properties and the Quaker Oats Co. as well as the E. I. du Pont de Nemours Co., Inc. for providing much useful information about both tetrahydrofuran and poly(tetrahydrofuran). Permission from copyright owners to reproduce material originally appearing in their publications is gratefully acknowledged. These include the Americal Chemical Society, publishers of *Macromolecules* and *Advances in Chemistry*; the E. I. du Pont de Nemours Co., Inc.; Gakujutsu Bunken Fukyukai of the Tokyo Institute of Technology, publishers of *Reports on Progress in Polymer Physics in Japan*; Hüthig and Wepf Verlag, publishers of *Die Makromolekuläre Chemie*; IPC Science and Technology Press, Ltd., publishers of *Polymer*; Marcel Dekker, publishers of the *Journal of Macromolecular Science*; Pergamon Press, Inc., publishers of *European Polymer Journal* and *Polymer Science USSR*; the Rubber Division of the American Chemical Society, publishers of *Rubber Chemistry and Technology*; The Quaker Oats Co.; the Society of Chemical Industry, publishers of *The British Polymer Journal*; Springer–Verlag, publishers of *Advances in Polymer Science*; and John Wiley and Sons, Inc., publishers of *Journal of Applied Polymer Science* and *Journal of Polymer Science*. Finally, I thank M. B. Huglin, editor, for inviting a Dreyfuss to write a monograph on poly(tetrahydrofuran); Gordon and Breach, the publishers, for their patience through repeated delays in delivery of the manuscript; and my readers, without whom there would have been no need to write this monograph.

P. DREYFUSS

CHAPTER 1

Introduction

1.1 ONE MONOMER, ONE POLYMER, MANY NAMES

The formulae for the monomer and the polymer featured in this monograph are simple and unique. The monomer is a cyclic ether containing four methylene groups connected by an oxygen bridge. It is planar with some internal strain resulting from repulsion of the eclipsed hydrogen atoms.[1,2] Normally this monomer is stable but, under the right conditions, ring-opening polymerization occurs and a linear polyether forms. No polymeric isomers are possible.

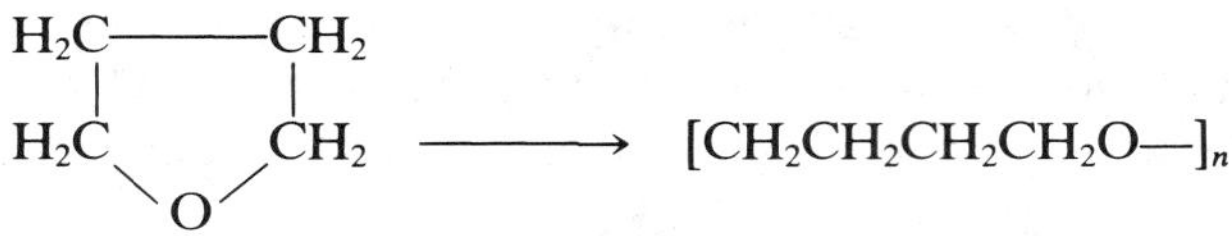

The monomer is listed in Chemical Abstracts under the entry "furan, tetrahydro" with a registry number of 109-99-9. The general polymer is currently indexed either under "furan, tetrahydro, polymers, homopolymer" with registry number 24979-97-3 and formula $(C_4H_8O)_x$ or under "poly(oxy-1,4-butanediyl)" with registry number 26913-43-9 and formula $(C_4H_8O)_n$. An inquiry to Chemical Abstracts revealed that formulas with subscript x are monomer based whereas formulas with the subscript n are given in terms of the polymer repeat unit. Moreover, there is a different entry for the α,ω-glycol under the name "poly(oxy-1,4-butanediyl), α-hydro-ω-hydroxy" with registry number 25190–06–1 and formula $(C_4H_8O)_n H_2O$. Polymers with other known endgroups have still other registry numbers. References under the various registry numbers and names are not necessarily crossindexed, so the literature searcher is obliged to check a variety of entries.

Throughout this monograph the common names tetrahydrofuran (THF) and poly(tetrahydrofuran) (PTHF) will be used. Alternative names, which appear in the literature, will only be mentioned here. For the monomer these include tetramethylene oxide, oxacyclopentane, 1,4-epoxybutane, diethylene oxide, oxolane, and H_4 furan. Alternative names for the polymer include poly (tetramethylene oxide) and sometimes poly(oxytetramethylene) or poly(tetrahydrofurane).

1.2 HISTORICAL

Polymerization of THF was first observed in Germany in the late 1930's by Meerwein and his coworkers[3] and was studied extensively by them throughout the 1940s. However, the work did not generally become known until after World War II and even then it was available only in the form of microfilmed reports.[4] Not until 1960, with the publication of Meerwein's review,[5] did the scope of this excellent work become generally available.

Meanwhile, interest in PTHF has increased steadily throughout the world, many other groups of investigators have entered the field, hundreds of papers have been published and applications have been found. Although PTHF has been examined as a possible general-purpose rubber, it has continued to be too expensive for such applications, and finds utility mainly as the soft segment in certain block copolymers to be described below. Summaries of various aspects of the work in THF polymerization have appeared regularly.[6–10] They have generally been limited to a single topic such as kinetics, or dealt with the work from a specific laboratory such as that of Meerwein. In this short work an attempt is made to give an overview of THF polymerizations and of the properties of the resultant polymer. Greater emphasis will be placed on advances that have occurred during the past 10 years, since earlier information is available elsewhere.

REFERENCES

1. "Du Pont Tetrahydrofuran as a Chemical Intermediate", Electrochemicals Department, E.I. du Pont de Nemours and Co., Inc., Wilmington, Del. (undated company publication).
2. H. K. Hall, *Polymer Preprints*, **6**, 535 (1965).
3. H. Meerwein, German Patent 741, 476 (June 21, 1939).
4. US Office of Technical Service Reports, Dept. of Commerce, as enumerated in Refs 1–15 of A. C. Farthing, "Polymers from 1,3- and Higher Epoxides" in "High Polymers", Vol. XIII, N. G. Gaylord (ed.), Interscience, New York, NY Chap. 5, p. 310 (1963).
5. H. Meerwein, D. Delfs and H. Morshel, *Angew. Chem.*, **72**, 927 (1960).
6. J. Furukawa and T. Saegusa, "Polymerization of Aldehydes and Oxides", Interscience, New York, NY (1963).
7. P. Dreyfuss and M. P. Dreyfuss, *Adv. Polym. Sci.*, **4**, 528 (1967).
8. P. Dreyfuss and M. P. Dreyfuss, in "Encyclopedia of Polymer Science and Technology", Vol. 13, H. Mark and N. G. Gaylord (eds.), Wiley, New York, NY p. 670 (1970).
9. T. Saegusa and S. Kobayashi, *Prog. Polym. Sci.*, (Japan), **6**, 107 (1973).
10. P. Dreyfuss and M. P. Dreyfuss, in "Comprehensive Chemical Kinetics", Vol. 15, C. H. Bamford and C. F. H. Tipper (eds.), Elsevier, Amsterdam-W., Netherlands, Chap. 4, p. 259 (1976).

CHAPTER 2

The Monomer

2.1 SYNTHESIS

2.1.1 Furfural route

The oldest and, for a long time, the only industrial synthetic route to THF is based on furfural as feedstock.[1-5] This method is still being used in the US by the Quaker Oats Co., which

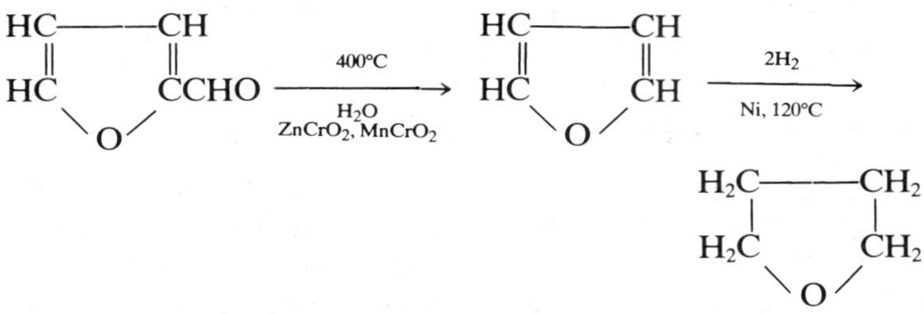

produces large quantities of furfural. The raw materials for this process include bagasse (the residue from sugar-cane or sugar-beet extraction), corn cobs, oat hulls, rice hulls and cottonseed hulls. Worldwide, the furfural route has been largely replaced by other synthetic routes based on acetylene/formaldehyde, maleic anhydride and butadiene as feedstocks. This situation may change as petroleum supplies decrease, petroleum prices increase and manufacturers return to annually renewable agricultural sources as raw materials.

2.1.2 Reppe process

The process based on acetylene/formaldehyde, the Reppe route, was originally developed by Germany during World War II to make butadiene.[6,7] After the War, Badische Analin-

und Soda-Fabrik AG (BASF) in Germany omitted the last step of the process and used it to produce THF. The company

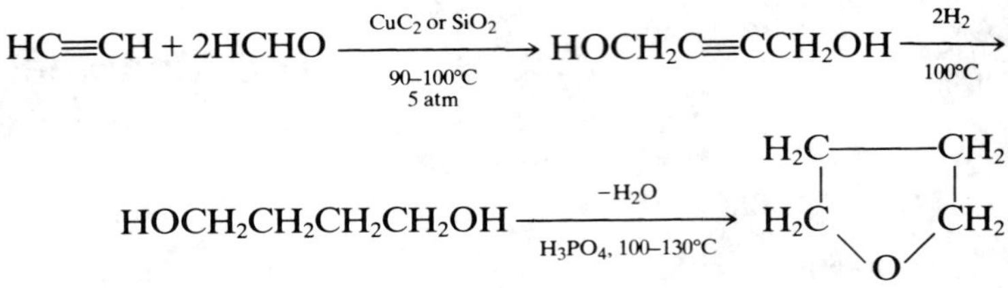

claims that the Reppe route is the most economical one for large-scale production with greater than 5,000 tons/year capacity. DuPont, the only US producer of THF besides Quaker Oats, also uses this process[1–3,6,7] or a similar one which is also based on acetylene and formaldehyde.[8]

2.1.3 Maleic-anhydride route

Mitsubishi Petrochemical of Japan has developed a process based on the direct catalytic hydrogenation of maleic anhydride.[6,7] The choice of catalyst is very important. The nature of the commercial catalyst has not been revealed in the open literature but it probably contains nickel and a precious metal.

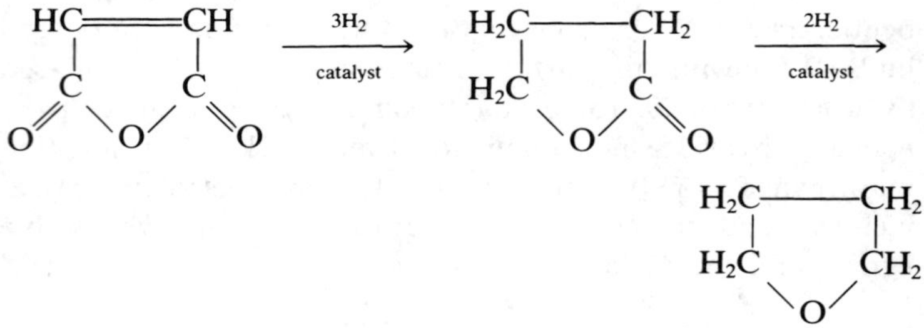

According to Mitsubishi, the Reppe process with its requirement for high pressure in the reactor is uneconomical on a smaller scale with a capacity of less than 5,000 tons/year. One of the key features of the route using maleic anhydride as feedstock is the fact that the ratio of the intermediate byproduct, γ-butyrolactone, to THF can be varied over wide limits. This is useful because γ-butyrolactone is a selective solvent for acetylene and is an important intermediate for the production of polyvinylpyrrolidone.[6] This process is being widely used.

2.1.4 1,4-Dichlorobutene route

The newest route with industrial potential, developed by Toyo Soda Co., uses butadiene as feedstock and also gives chloroprene, a valuable monomer.[6]

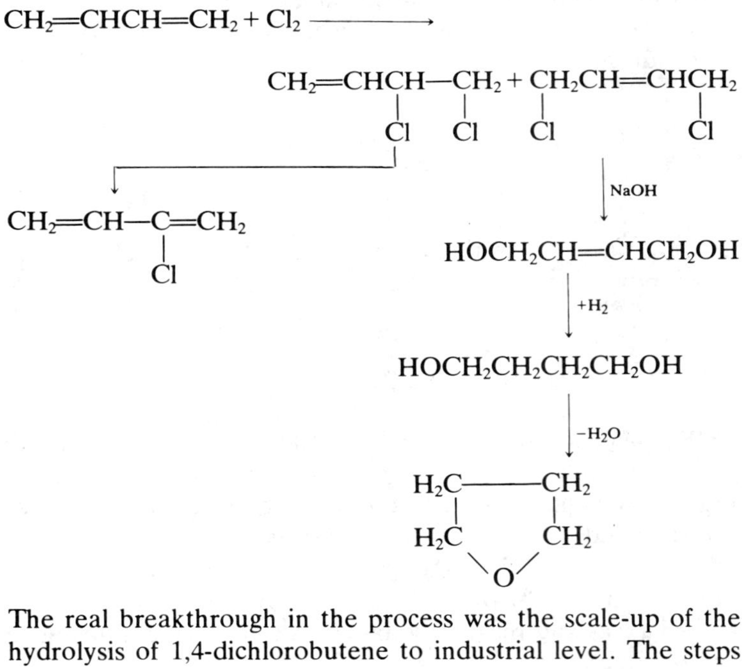

The real breakthrough in the process was the scale-up of the hydrolysis of 1,4-dichlorobutene to industrial level. The steps from 1,4-butenediol are similar to those in the Reppe route.

2.1.5 Comments

Each of these processes involves many steps, so the price of THF remains high relative to that of monomers like butadiene, ethylene and propylene, and too high for PTHF to complete as a general-purpose rubber. But as a solvent, especially for such high-volume uses as cast-vinyl film or in vinyl top-coating in automobile seat covers, THF has few competitors. Thus for speciality applications, a continuous supply of monomer for PTHF production should continue to be available. If, and when, a more economical route to the production of THF—perhaps a one-step hydration and cyclization of butadiene—is found, this economic picture could change.

There is a patent claiming direct preparation of THF by codimerization of ethylene oxide and ethylene in the presence of HI as promoter and catalytic amounts of Group VIII noble transition-metal compounds such as $Ru(CO)_3[(C_6H_5)_3]_2$.[10] Depending on temperature and pressure, conversion to

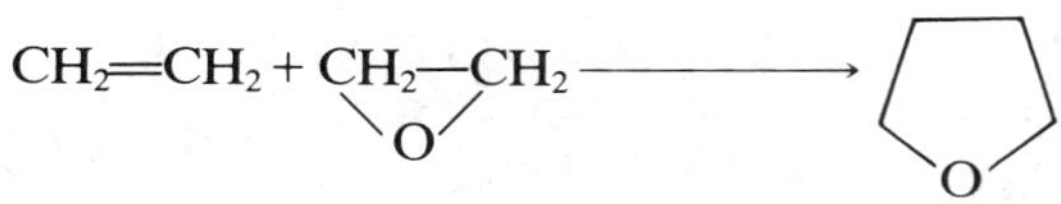

product was often in the range 40–84%, but the yield of THF was rarely more than 30 mole % of the total yield.[10]

2.2 PHYSICAL PROPERTIES

THF is an industrial solvent with the very useful properties shown in Table 2.1.[11,12] It is miscible with water and common organic solvents. The effect of temperature on the vapor pressure and on the density of THF is shown in Figures 2.1 and 2.2, respectively. Note that at ordinary temperatures a vapor is given off. Since THF is a flammable liquid, a fire hazard exists and reasonable precautions and care in handling are necessary.

THF requires only the normal precautions recommended for handling volatile organic compounds. It should be used only in suitable equipment and with adequate ventilation. Goggles should be worn to avoid getting THF in the eyes. THF can be stored and handled in ordinary steel tanks and piping. However, since THF is an excellent solvent, welded piping is

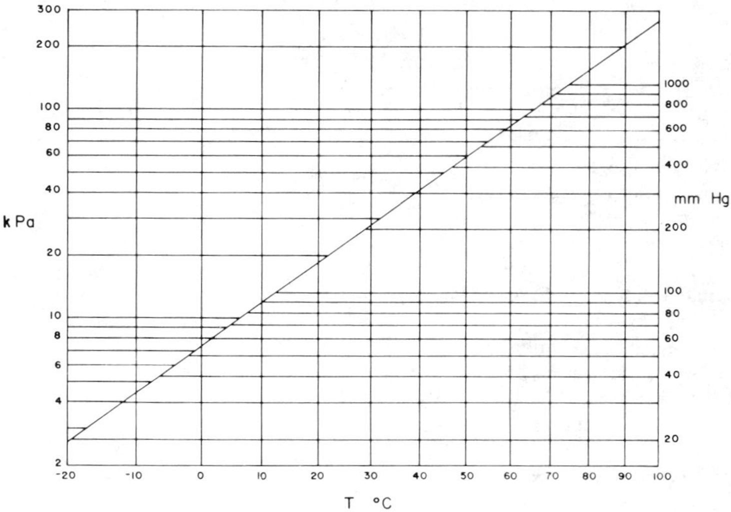

FIGURE 2.1 Vapor pressure of THF.[11]

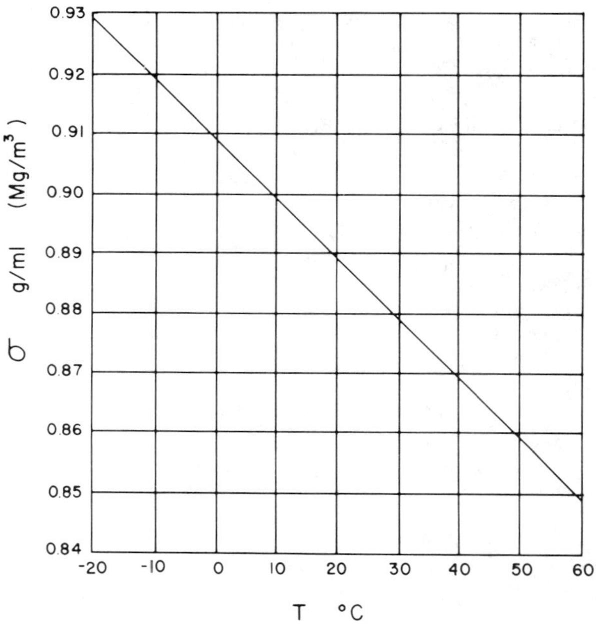

FIGURE 2.2 Density, σ, of THF.[11]

TABLE 2.1
Physical Properties of Tetrahydrofuran[11,12]

Property	Value
Molecular weight	72.108
Boiling point (at 760 mm Hg), °C	66
Freezing point, °C	−108.5
Vapor pressure, 20°C, mm Hg	143
kPa	19.1
Density—liquid, 20°C, g/ml (Mg/m^3)	0.888
vapor (air = 1)	2.49
Viscosity, 20°C, cP (mPa · s)	0.48
Surface tension in air, 25°C, dyn/cm (mN/m)	26.4
Refractive index, n_D^{20}	1.4073
Heat of vaporization (at bp), cal/g	95
kJ/kg	397
Heat of combustion, kcal/mol	597
MJ/kg	34.6
Specific heat, liquid,	
20°C, cal/g · °C	0.469
kJ/kg K	1.962
50°C, cal/g · °C	0.496
kJ/kg K	2.090
vapor, 66°C, cal/g · °C	0.37
kJ/kg K	1.55
Coefficient of thermal expansion,	
10–20°C, °C^{-1} (average)	0.00126
Flash point (TCC), °C	−14.4
Autoignition temperature, °C	321
Flammability limits in air,	
25°C, lower, %vol. THF	2
upper, %vol. THF	11.8
Critical temperature, °C	268
Critical pressure, atm	51.2
MPa	5.19
Dielectric constant, ϵ, 20°C	7.54
30°C	7.25
Conductivity, 25°C, μmhos/cm	0.015
μS/m	1.5
Dipole moment, μ, 25–50°C,	
Debye units	1.7
Solubility parameter, δ, (cal/cm^3)$^{1/2}$	9.7
(kJ/dm^3)$^{1/2}$	19.8
Hydrogen bonding index, γ	2.7

TABLE 2.1 (*continued*)

Property	Value
Evaporation rate (n-butyl acetate = 1)	8.0
Miscibility: water, esters, ketones, alcohols, diethyl ether; aliphatic, and chlorinated hydrocarbons	infinite
Explosion limits (% by volume in air, 25°C)	1.8–11.8
Threshold limit value, ppm	200

better than screwed fittings, and pump and valve packing must be tight and nonsoluble.[12]

2.3 CHEMICAL PROPERTIES

THF is a saturated cyclic ether and many of its chemical properties are similar to those of aliphatic ethers. Thus, THF is inert to such chemicals as sodium or potassium, dilute mineral acids, Grignard reagents, oxidizing agents and metal hydrides.[12,13] As in other organic ethers, peroxide can form in unstabilized THF. Therefore, a nonvolatile antioxidant such as 2,6-di-*t*-butyl-4-methylphenol (butylated hydroxytoluene) is generally added to prevent the formation or accumulation of significant quantities of THF peroxide during storage and transfer. However, unlike linear aliphatic ethers, THF can undergo a wide variety of chemical reactions including polymerization, oxygen-replacement, ring-substitution, hydrolytic-cleavage and molecular-addition reactions. One of these reactions, its polymerization, will be discussed in detail in this monograph.

2.4 PHYSIOLOGICAL PROPERTIES

THF is a volatile, flammable liquid, and may be hazardous to health if not handled properly. From the standpoint of toxicity,

THF is about as dangerous as methyl ethyl ketone. Prolonged exposure to THF vapor may produce narcosis, nausea, dizziness and headache. However, these symptoms disappear after moving to a fresh-air atmosphere. THF defats the skin, which may lead to irritation, and it is an "eye irritant" as defined by the US Federal Hazardous Substances Act. Adequate ventilation and protective clothing used with most common organic solvents are necessary for the safe handling of THF. Gloves made of poly(vinyl alcohol) are recommended protection for the hands.[13]

Physiological studies conducted on du Pont THF by its Haskele Laboratory for Toxicology and Industrial Medicine have shown no demonstrable pathology in the lungs, heart, liver, kidney or other organs of test animals exposed to high concentrations of THF vapor in air. Other tests at the laboratories of BASF, Germany, led to the same conclusions.[11] According to the du Pont studies, reports in the literature that THF itself produces damage to vital organs are inaccurate. The reported damage is real, but was caused by considerable amounts of furan impurity in the THF tested.

The US Department of Labor has ruled that an employee's exposure to THF shall not exceed 200 ppm. The threshold limit value (TLV) of 200 ppm, specified by the Occupational Safety and Health Administration of the USA, is in agreement. Such an exposure unknowingly is highly unlikely, because most people can detect THF at a concentration of 25–50 ppm.

THF should not be ingested. If THF is swallowed, vomiting should *not* be induced because of the danger of aspiration of the solvent into the lungs, where it can produce severe lung irritation.

2.5 PURIFICATION

As stated above, THF may form THF peroxide, really 2-hydroperoxy THF, on exposure to air. Even though this hydroperoxide tends to decompose quietly when heated to

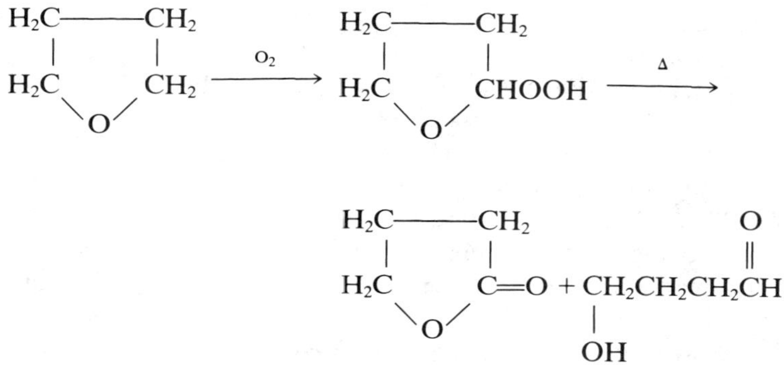

give γ-butyrolactone and 4-hydroxybutyraldehyde,[14] distillations involving THF should *never* be continued to dryness. When concentrated, THF peroxide is transformed to a true cyclic peroxide, thought to be a trimer, which is unstable to isolation and decomposes violently.[14] If there is any suspicion that more than a trace of THF peroxide is present in a sample, a test for THF peroxide content should be carried out. Qualitatively this can be done with KI-starch paper[15] or acidic aqueous iodide solution.[16] For a quantitative determination du Pont and Quaker Oats recommend the following as a simple accurate method for determination of THF peroxide:[11,12]

1) place 100 ml of distilled water in each of two Erlenmeyer flasks;

2) to each flask add about five drops of 5% ammonium molybdate solution, 25 ml of a 1:4 mixture of concentrate sulfuric acid and water, then 25 ml of a 10% potassium iodide solution;

3) with a pipette, add a 25 ml sample of THF to one flask, stopper both flasks, shake and place in the dark for 15–20 minutes;

4) titrate the contents of each flask with 0.02 N sodium thiosulfate until the solutions become colorless;

5) calculate

$$\%\text{THF-peroxide} = (\text{ml thio in sample} - \text{ml thio in blank}) \times \text{Normality of thio} \times 0.23$$

If peroxide in excess of 0.05 wt.% is present, the THF should be discarded or carefully treated to remove peroxide before being used. A preliminary small-scale test on 1–5 ml is recommended. Traces of peroxide can be removed by treatment with cuprous chloride.[16] Alternatively, THF peroxide can be destroyed by agitating the THF with flake sodium hydroxide, using 5 g of caustic soda for each 100 g of solvent or a 73% aqueous solution of caustic soda. Treatment of THF containing more than 0.05 wt.% peroxide should be carried out by incremental addition of sodium hydroxide to avoid a runaway reaction.[11]

THF which has been recovered and stored for re-use should be restabilized or stored under a blanket of dry nitrogen gas. Molecular sieves are useful for removing water from THF–organic-solvent mixtures. Moisture contents as low as 0.002 wt.% have been attained by this procedure.[17]

If THF is to be polymerized, it needs further purification and thorough drying. The best laboratory procedure seems to be distillation under nitrogen from lithium aluminum hydride.[16] Refluxing over and distillation from sodium dispersion or potassium metal under nitrogen is also frequently used. Generally, especially for high-vacuum studies, THF is finally distilled under nitrogen or vacuum from blue sodium–benzophenone or green sodium–anthracene or green sodium–naphthalene complex. Industrially, distillation under applied pressure can be used to reduce the water content of THF to about 0.5 wt.%.[12,17,18] Atmospheric distillation alone is not suitable, because THF forms an azeotrope containing 5.3 wt.% water.

Dehydrating with calcium hydride has been reported to be a safe method for purifying THF. However, a violent explosion occurred recently in a university laboratory during refluxing with calcium hydride.[15] The cause of the explosion is not known, but it is postulated that cleavage of the THF by reaction with excess hydride occurred due to localized overheating on the wall of the reaction vessel.[14,15]

This section can only be concluded by quoting John S. Coates, du Pont spokesman:[14] "In this laboratory, we have found THF to be an extremely useful solvent, and, when handled properly, poses no more significant safety hazards than other organic solvents." It may be added, in the context of the present monograph, that an interesting and useful polymer can be prepared from THF.

REFERENCES

1. A. C. Farthing, "Polymers from 1,3- and Higher Epoxides" in "High Polymers", Vol. XIII, N. G. Gaylord (ed.), Interscience, New York, NY, Chap. 5 (1963).
2. D. Starr and R. M. Hixon, in "Organic Syntheses", Coll. Vol. 2, Wiley, New York, NY, p. 566 (1963).
3. K. Matsuda, *Chemtech.*, **4**, 744 (Dec. 1974).
4. C. R. Noller, "Chemistry of Organic Compounds", W. B. Saunders, Philadelphia, PA, pp. 586–7 (1951).
5. "The Story of Furfural and the Chemicals Division of the Quaker Oats Company", Chemicals Division, The Quaker Oats Company, Chicago, Ill. 60654 (undated).
6. T. Yoshimura, *Chem. Eng.*, **76**, 70 (Aug. 11, 1969).
7. S. Ushio, *Chem. Eng.*, **78**, 24 (Dec. 27, 1971).
8. *Chem. Eng. News*, **44**(50), 29 (Dec. 5, 1966).
9. J. Kanetaka, T. Asano and S. Masamune, *Ind. and Eng. Chem.*, **62**(4), 24 (1970).
10. V. P. Kurkov, US Patent 3,933,861 (to Chevron Research Co.) (Jan. 20, 1976).
11. "Tetrahydrofuran, Properties, Uses, Storage and Handling", E. I. du Pont de Nemours and Co., Wilmington, Del. 19898 (undated).
12. "QO Tetrahydrofuran", Bulletin 148A, Chemicals Division, The Quaker Oats Co., Chicago, Ill. 60654 (undated).
13. "Du Pont Tetrahydrofuran as a Chemical Intermediate", Electrochemicals Department, E. I. du Pont de Nemours and Co., Wilmington, Del. 19898 (undated).
14. J. S. Coates, *Chem. Eng. News*, **56**(6), 3 (Feb. 6, 1978).
15. R. Holanbrenner and W. Kläui, *Chem. Eng. News*, **56**(6), 3 (Feb. 6, 1978).
16. E. J. Corey (ed.), *Organic Syntheses*, Vol. 46, Wiley, New York, NY, p. 105 (1966).

17. "Recovery of THF", Electrochemicals Department, E. I. du Pont de Nemours and Co., Wilmington, Del. 19898 (undated).
18. "Properties and Uses of THF", Electrochemicals Department, E. I. du Pont de Nemours and Co., Wilmington, Del. 19898 (undated).

CHAPTER 3

Polymerization

3.1 INTRODUCTION

The THF molecule is a planar five-membered ring with some internal strain resulting from repulsion of eclipsed hydrogen atoms. Since the ring contains an oxygen atom with two unshared pairs of electrons, THF is a nucleophilic monomer where steric interference to potential electron acceptors is low. A slightly negative free energy of polymerization of about −3350 J/mol at 25°C[1] has been estimated from the experimentally derived heat of polymerization of −18.8 kJ/mol and an entropy of polymerization of about $-75.3\ \mathrm{J\ deg^{-1}\ mol^{-1}}$. Thus, under the right conditions polymerization can occur. Purity of reagents, dryness of apparatus, choice of initiator, atmosphere, monomer concentration and solvent will determine the amount, and many properties, of the polymer prepared.

3.1.1 The counterion

THF polymerizes only by a cationic ring-opening mechanism.[1–6] The propagating species in THF polymerizations is a tertiary oxonium ion:

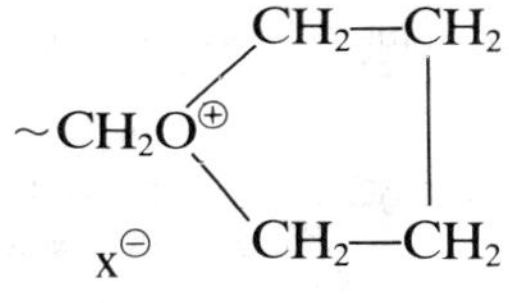

and must, for reasons of electrical neutrality, be associated with a negatively charged species, the counter ion X^-. (At this stage, the possibility that oxonium-ion–anion complexes from super acids can collapse to give esters will not be discussed. This will feature prominently in the kinetics chapter.) The early work of Meerwein and coworkers has shown that polymerization continues only if the propagation rate is faster than the rate of irreversible reaction of the oxonium ion with its

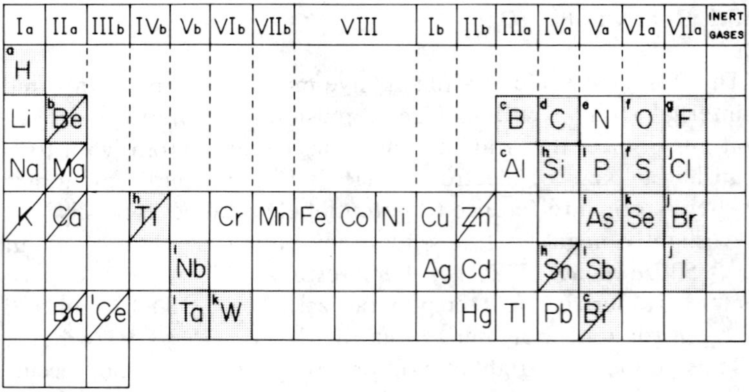

FIGURE 3.1 Periodic chart of elements used in initiation of polymerizations of THF. The elements in open squares were used as perchlorate salts.[8] Those in shaded squares were taken from the literature. A diagonal across the square indicates that, although the element was tried, polymerization was not initiated.

[a]As H^+ with a variety of counterions such as $ClO_4^{\ominus}$, $PF_6^{\ominus}$, $SO_3F^{\ominus}$, . . .

[b]As $BeCl_2$.[3]

[c]As BCl_3, BF_3, $AlCl_3$, $AlBr_3$, AlR_3, $BiCl_3$, $Al(ClO_4)_3$.[1,3,8]

[d]As $R_3C^{\oplus}$, $RCO^{\oplus}$, $RC(OR')_2^{\oplus}$.[13]

[e]As $NR_4^{\oplus}$,[9–12] $NO^{\oplus}$, $NO_2^{\oplus}$.[13]

[f]As $R_3O^{\oplus}$, $R_3S^{\oplus}$ or $Ar_3S^{\oplus}$;[1,3,14,15] also in numerous anions.

[g]As $F^{\ominus}$, very common as nonmetallic element in complex anions; as RF in conjugation with metal salts.

[h]As SiF_4, $SnCl_4$, $TiCl_4$.[3]

[i]As PF_5, AsF_5, SbF_5, $SbCl_5$, $NbCl_5$, $TaCl_5$, NH_3, RNH_2, PR_3, . . .[1,16–18]

[j]As $(R)R'X^{\oplus}$;[19–22] as $Cl^{\ominus}$, $Br^{\ominus}$, $I^{\ominus}$ in complex anions; $ClO_4^{\ominus}$; as RX in conjunction with metal salts.

[k]As SeF_6 or WCl_6 or $Ar_2Se^{\oplus}$.[16,23–25]

[l]Ce is element 58 and the first member of the lanthanide series.

In all cases R, R′ = alkyl or H: Ar = aryl; X = Cl, Br, I.

accompanying anion.[7] Simple anions like Cl^- are not suitable counterions because the cation–anion complex is unstable with respect to THF and alkyl halides. Stable counterions are complex ions including anions like PF_6^-, AsF_6^-, SbF_6^-, $SbCl_6^-$, BF_4^-, $SO_3CF_3^-$, SO_3F^-, ClO_4^-. A complete discussion of counterion chemistry in THF polymerizations is not appropriate in this introduction. However, a few general comments should be made here; these will be detailed later.

It is very useful to consider these anions in terms of the position of the elements of which they are composed in the periodic table (see Figure 3.1). As already stated, oxonium ions are stabilized complex ions. Therefore, the elements must be a component of a complex ion. In addition, the elements that comprise stable counterions for THF polymerizations are found almost exclusively among the nonmetals in groups VA, VIA and VIIa. These elements include the most highly electronegative element, F, and are electron acceptors. More firmly bonded elements in a complex ion lead to more stable and more suitable counterions for THF polymerizations and fewer side reactions. Thus, counter ions with F will be more stable than those with Cl or Br, but SbF_6^- is potentially more stable than PF_6^-. Polymerizations can occur in the presence of complex anions whose central atom is derived from elements in other groups such as BF_4^- from boron in group IIIA and $FeCl_4^-$ from iron in group VIII; but side reactions increase with anions containing metallic components from elements further to the left in the periodic table.

3.1.2 Reagents and apparatus

Since THF polymerizes only by an ionic mechanism, requirements for a successful polymerization include: pure, dry reagents; dry apparatus; and exclusion of air by the use of either high vacuum or an inert atmosphere such as dry nitrogen. Methods for purifying the monomer have already been described in Chapter 2. Solvents are similarly purified, and calcium hydride is the drying agent most frequently used in the laboratory.

3.1.3 Temperature, Solvent

A suitable temperature and limited solvent, depending on the temperature, are also critical factors, because the polymerization of THF is an equilibrium process. Hence, for every temperature there is an equilibrium monomer concentration, M_e, that is thermodynamically determined. There is also a ceiling temperature, T_c, above which no polymerization at all will occur. A plot of conversion to PTHF as a function of temperature is shown in Figure 3.2.[26] From these data the value derived for T_c is 83 ± 2°C. Dainton and Ivin[27] have related M_e to the absolute temperature T in the equation:

$$\ln [M]_e = \frac{\Delta H_p}{RT} - \frac{\Delta S_p^0}{R} = \frac{\Delta G_p}{RT},$$

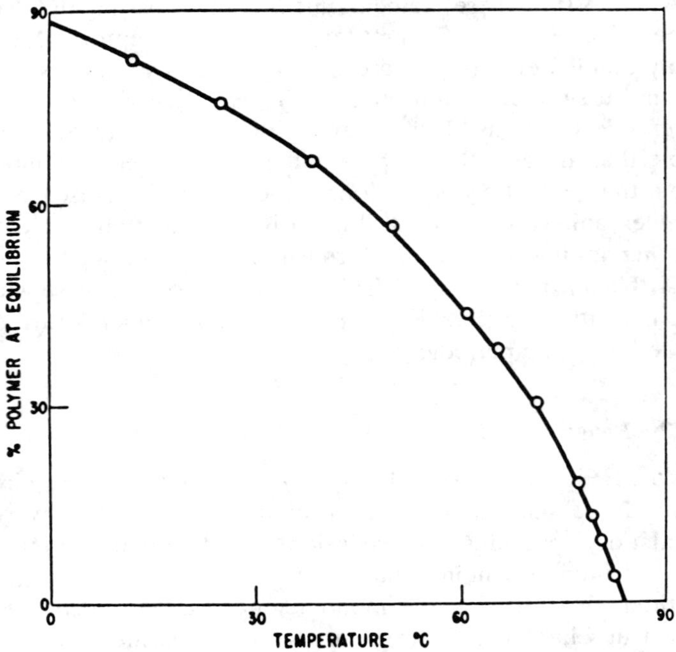

FIGURE 3.2 Effect of temperature on conversion to polytetrahydrofuran in bulk tetrahydrofuran polymerization.[1,26]

where ΔH_p is the heat of polymerization under the prevailing experimental conditions, ΔS_p^0 is the entropy change for $[M]_e = 1$ M, and ΔG_p is the free energy of polymerization at temperature T and normal pressure.

(Some readers may be interested to know that this classic derivation of Dainton and Ivin turns out to be a special case in a 1972 theoretical treatment of the reaction kinetics of reversible polymerizations.[28] The thermodynamics of equilibrium polymerizations have recently been reviewed by Sawada.[29])

Figure 3.3 shows that for THF polymerizations the plot of ln $[M]_e$ versus $1/T$ gives a straight line. The values of ΔH_p and ΔS_p^0 obtained are, respectively, -18.0 to -23.4 kJ mol^{-1} and -71 to -87 J deg^{-1} mol^{-1}.[1] Several investigators have analyzed THF polymerizations in this way, some with refinements to allow for nonideal mixing, etc.[30–31]

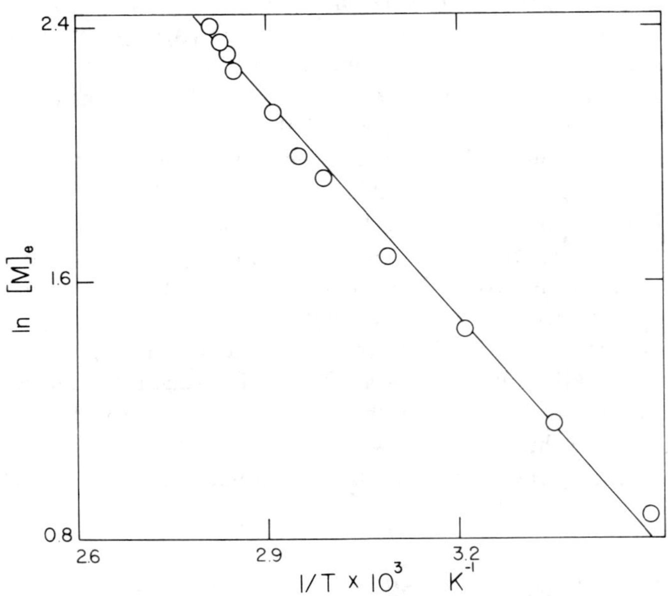

FIGURE 3.3 Logarithm of equilibrium THF concentration, ln $[M]_e$, as a function of reciprocal temperature plotted according to the equation of Dainton and Ivin.[26,27] See page 20.

Penczek and coworkers[32] have shown that both the enthalpy and entropy of polymerization are somewhat dependent on both counterion and solvent. For the $CF_3SO_3^{\ominus}$ counter ion, for example, ΔH_p and ΔS_p^0 are respectively equal to: -25.1 ± 1.3 kJ mol^{-1} and -104.6 ± 4.2 J mol^{-1} deg^{-1} in CCl_4; and -23.4 ± 0.8 kJ mol^{-1} and -82.4 ± 3.3 J mol^{-1} deg^{-1} in CH_2Cl_2. The reasons for these differences are discussed in Penczek's paper and are related to structural factors and solvation.

Other methods have resulted in values of ΔH_p from -7.5 to -38.1 kJ mol^{-1} and of ΔS_p from -16.3 to -75.3 J deg^{-1} mol^{-1}.

Since the conversion to PTHF at any given temperature clearly depends on the monomer concentration and is independent of the polymer concentration, inert diluent reduces the conversion to polymer at a given temperature. If a solvent is used, the $[M]_e$ characteristic of a given temperature and based on the whole system can be achieved only at the expense of polymer conversion. In THF polymerizations carried to equilibrium conversion at 30°C, polymerization in bulk gives 72% PTHF, while polymerization in a solution containing 62.5 vol.% THF and 37.5 vol.% methylene chloride gives 27% PTHF. Largely because of these considerations, most polymerizations of THF are carried out at room temperature and below, especially if an inert diluent is used.

There are practical limits to how low a temperature can be used. Below about −20°C, the rate of propagation becomes so slow that prohibitively long times are required for significant conversions to PTHF to be achieved.

Temperature also affects the side reactions that sometimes occur in THF polymerizations. These side reactions can occur during initiation, propagation or termination, and will be considered in other sections, where their effects on conversion, mechanisms, end groups, molecular weight, etc., will be pointed out.

3.1.4 Additional comments

In spite of all these considerations, very well-defined PTHF polymers can be readily prepared. "Living" oxonium-ion polymerizations, i.e., polymerizations which are, for all practical purposes, free from transfer and termination reactions, are possible. This means that PTHF of any desired molecular weight with a narrow molecular-weight distribution and controlled end groups can be prepared. However, a careful balancing of thermodynamic, kinetic and mechanistic considerations is essential if this goal is to be achieved.

3.2 MECHANISM

3.2.1 Initiation

Introduction A very large number of materials have been used to initiate THF polymerization and many more have been tried without success. While this monograph was being written, a study was made in our laboratory of the initiating capabilities of 22 different metal perchlorates when added to organic halides in THF. Activity of the metal perchlorates seemed to vary in a systematic way with the position of the metals in the periodic table.[8] This led to a consideration of whether the behavior of all the materials used to initiate THF polymerization could be related to the position of their central elements in the periodic table. In my opinion, it can, and throughout this section relative reactivity is discussed in these terms. This means that we now have a tool for:

a) predicting the behavior of the remaining untested materials;

b) gaining new insight into hitherto unsuspected side reactions that might be occurring in the presence of some materials.

c) selecting an unreactive material, if that is what is needed in a particular application.

For those who prefer a more traditional approach to initiation mechanisms based only on the type of compound being used, this section has been subdivided according to the mechanism by which the required tertiary oxonium ion is formed.

Figure 3.1 shows the location in the periodic table of elements in compounds used for the initiation of THF polymerization. The footnotes describe the types of compound which have been tested for each of the elements. Their use in anions and the kind of anions that are useful for stabilizing the THF tertiary oxonium ion have already been discussed. This section deals primarily with reactions that ultimately lead to the THF tertiary oxonium ion required for THF propagation to polymer.

A suitable counterion may also be produced simultaneously, but most often a preformed counterion is assumed to accompany the positively charged species involved in the initiation process. Clearly, even if a mechanism of initiation exists for the positively charged species, but the counterion is one that reacts irreversibly with the THF oxonium ion to form a stable species, polymerization will not occur. Elements in our perchlorate study[8] are in open squares, while other elements taken from the literature and used in compounds for the initiation of THF polymerization are in shaded squares. A diagonal across the square indicates that, although at least one compound of the element was tried, no polymerization occurred. Some elements such as Li, B, Al, C and P have been used in a variety of compounds and, not surprisingly, with some elements more than one kind of reaction, depending on the compound, has been observed. These range from complete inertness to initiation by direct addition to THF or termination by addition to the growing center in a THF polymerization. Most elements are useful as initiators only in cations, i.e., they must be electron acceptors. Some can also function as electron donors but then termination of polymerization usually occurs. $BiCl_3$ or PF_3, for example, cannot be compared with PF_5. For this discussion, the same valence states and indeed the same type of compounds must be compared within groups. One cannot expect the chemical reactions of PF_5 to be the same as those of P_2O_5, but one can compare PF_5, SbF_5 and BF_3: all are

Lewis acids in the highest valence state characteristic of their group in the periodic table, and are electron acceptors. Relatively few elements (for example, F) are useful mainly in anions and a few elements (for example, O) can be used in both.

The chemistry of initiation of THF polymerization The manner in which compounds are used to induce polymerization of THF varies considerably. Since THF polymerizes only by an oxonium-ion mechanism,[1,3,7,33,34] as stated above, the basic requirement for polymerization is that a THF tertiary oxonium ion must be formed by some mechanism. Then, assuming that a suitable counterion is present, polymerization will follow.[31] The mechanisms by which the required tertiary oxonium ion is formed can be divided into several groups including:

1) direct acylation or alkylation of the oxygen of THF by "exchange" or addition;

2) *in situ* generation in THF from a variety of combinations such as epichlorohydrin or reactive halide and Lewis acid, reactive halide and metal salt, or sometimes Lewis acid alone;

3) strong protonic acid addition to the oxygen of THF to form an intermediate secondary oxonium ion, which is converted to a tertiary oxonium ion in a subsequent step; the protonic acid may be added directly or it may be formed *in situ*.

In the following discussion these mechanisms will be considered in some detail, and the types of compound which behave accordingly will be pointed out. Each group has been subdivided into subsections based on the type of compound used.

Group 1: Direct alkylation by "exchange" This process appears to occur only from oxygen, but the oxygen can be found in several different environments including tertiary oxonium salts, carboxonium salts and superacid esters. They all initiate in a similar manner, by a nucleophilic attack of the free electrons of the THF oxygen on the carbon adjacent to the oxygen.

Important tertiary oxonium salts include cyclic and acyclic trialkyloxonium ions:[1,31,34]

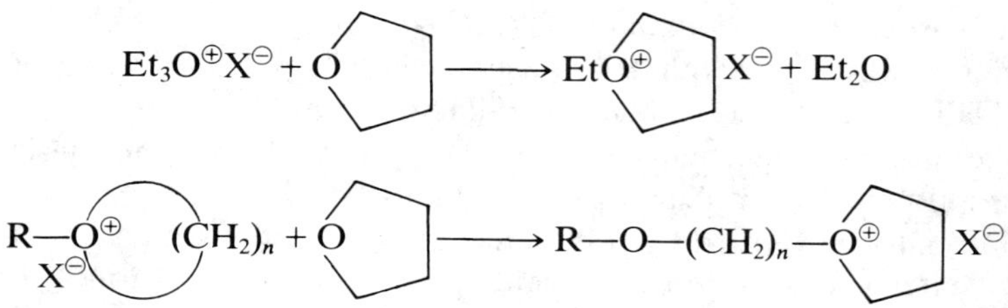

Carboxonium salts can also be acyclic[3] or cyclic, as in dioxolenium salts[35]

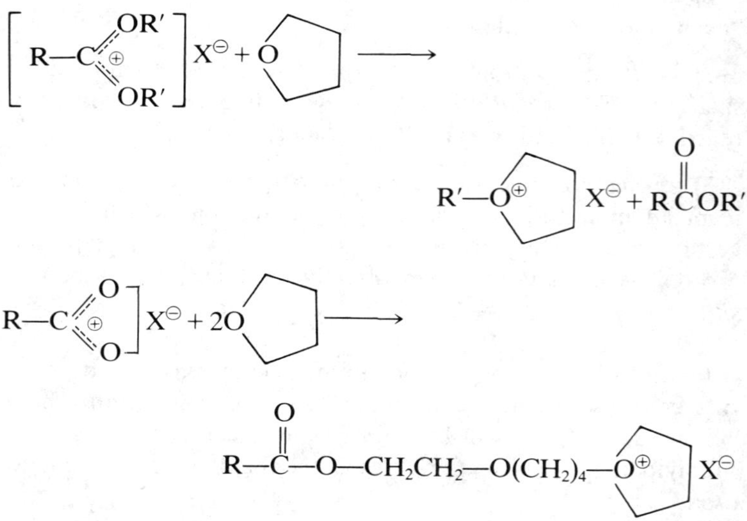

A variety of superacid esters (and anhydrides) are effective.[36] The reaction of methyltrifluoromethanesulfonate is an example.

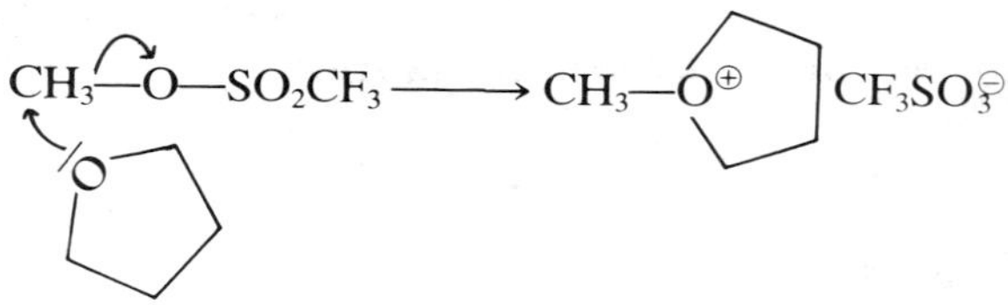

Polymerizations initiated by direct alkylation usually give clear colorless reaction mixtures.

Oxonium ions will react with cyclic sulfides to form sulfonium salts, but the reverse reaction cannot occur.[14] However, triarylsulfonium salts can function as photoinitiators of the cationic polymerization of THF.[15] The mechanism is not one of direct addition and its consideration is reserved for the discussion of group 3.

Group 1: Direct acylation or alkylation by addition This process occurs with acylium ions[34,37] and probably with $(C_6H_5)_2CH^{\oplus}$.[38]

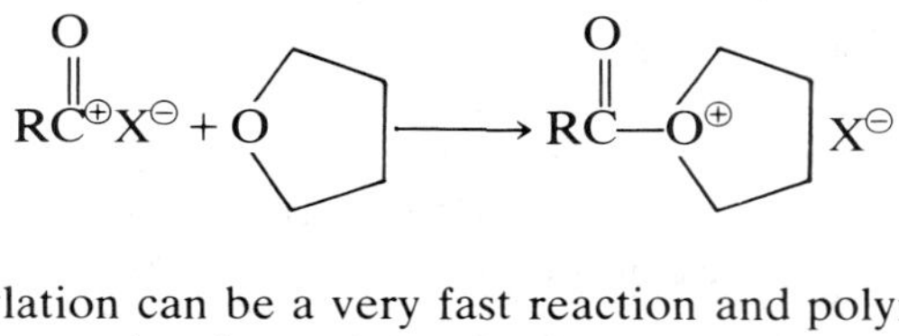

Direct acylation can be a very fast reaction and polymers with very narrow molecular weight distribution at low conversions can be formed.

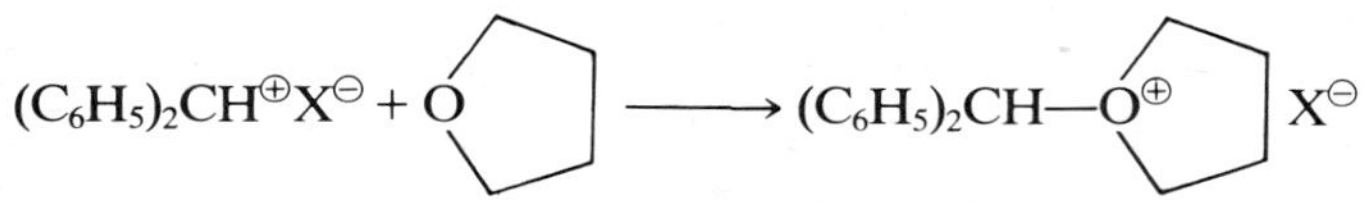

It would be wrong to assume that all carbenium ion salts with stable counterions will react in this way. Generally, initiation of polymerization does not occur; and if initiation does occur (see following description), the addition product is usually not formed.

Group 2: In situ formation of tertiary oxonium ion from

epichlorohydrin and a Lewis acid This process has been reported with BF_3, $SbCl_5$, $FeCl_3$ and $AlCl_3$.[3]

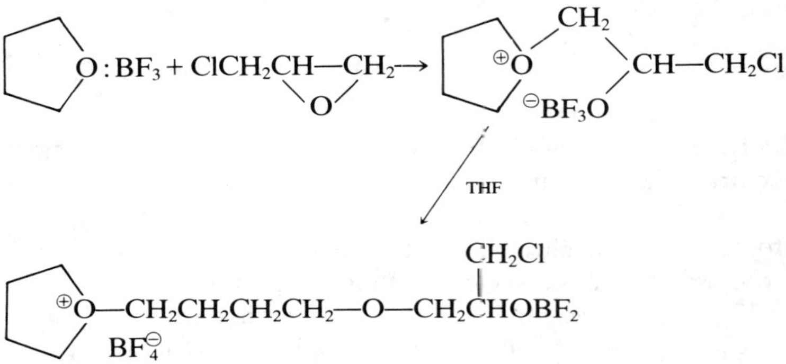

The $-OBF_2$ endgroup, or its equivalent if another Lewis acid is used, is not stable and undergoes further reaction. Unsubstituted, other monosubstituted and disubstituted epoxides react in the same way, but they are less efficient than epichlorohydrin.[39] Trisubstituted or tetrasubstituted epoxides are unreactive.[39,40] Compounds that react in this way were given the name "promoter" by Saegusa *et al.*[41] A promoter is a polymerizable or copolymerizable monomer, which acts irreversibly with the Lewis acid:THF complex and is incorporated into the chain.

Frequently, the trialkyloxonium salt is formed first using diethyl ether instead of THF, and then the polymer will be free of the colored products formed by undefined side reactions. Preformed trialkyl oxonium salts have been prepared in this way from BF_3, $SbCl_5$, $FeCl_3$, $AlCl_3$, PF_5, AsF_5 and SbF_5.[1,3] When *in situ* formation is carried out with PF_5, AsF_5 or SbF_5, it is probable that other reactions to be described below also occur simultaneously. Evidence for these side reactions was obtained by Dreyfuss and Kennedy[42] when they tried using PF_5 to form graft copolymers with polytetrahydrofuran branches from epoxidized elastomers. They found that considerable homopolymer formed along with the graft copolymer. With BF_3 the only product was graft copolymer. This is direct experimental evidence for the greater reactivity of group-VA pentavalent Lewis acids, since they can compete effectively in initiating THF polymerization by themselves, whereas the group-III Lewis acid, BF_3, requires a promoter.

Lewis acids including $BeCl_2$, $ZnCl_2$, BCl_3, $AlBr_3$, SiF_4, $TiCl_4$, $SbCl_3$ and $BiCl_3$ give alcoholates instead of trialkyl oxonium salts under the same conditions, and therefore do not initiate the polymerization of THF.[3]

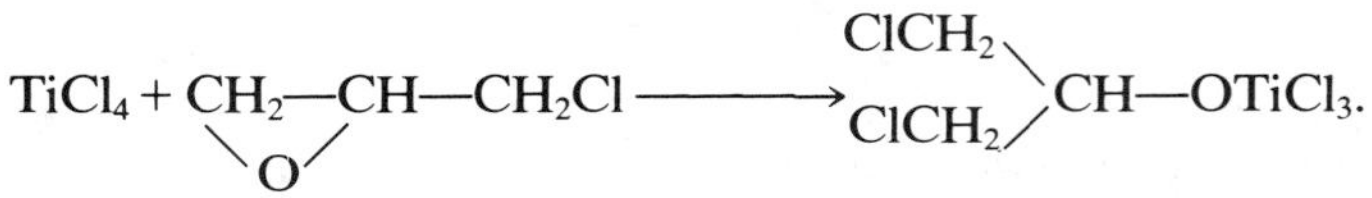

The reasons for the different behavior of the Lewis acids probably include differences in charge/radius ratio, in polarizability, in the ability to use empty d or f orbitals for back bonding and in lattice energy. A detailed discussion of these factors with relation to halides of boron, phosphorus and antimony has already appeared in the literature.[1] There it was noted that the same arguments apply also to iron and aluminum. In groups IIA and IIIB, where the metal halides are already quite stable and there is little tendency to form complex anions, it is not surprising that the *in situ* formation of THF tertiary oxonium ion does not occur. Halides from group IA were not tested but, for reasons given above, would not be expected to be reactive.

Group 2: In situ formation of tertiary oxonium ion from Lewis acid alone This process is most important with the pentavalent Lewis acids from elements in group VA. Only these Lewis acids are able to initiate THF polymerization without added epoxide.[1] Very stable oxonium ion–counterion complexes result, especially from fluorinated Lewis acids, and "living" poly(tetrahydrofuran) and poly(tetrahdyrofuran) of very high molecular weight can be prepared. About 10 years ago Dreyfuss and Dreyfuss[1] proposed that reaction proceeded in the same manner as the *in situ* formation of tertiary oxonium ion, where THF served the same function as epichlorohydrin. Hoene and Reichert recently showed by ^{31}P NMR that this suggestion is correct and that interactions continue until a dication is formed:[43]

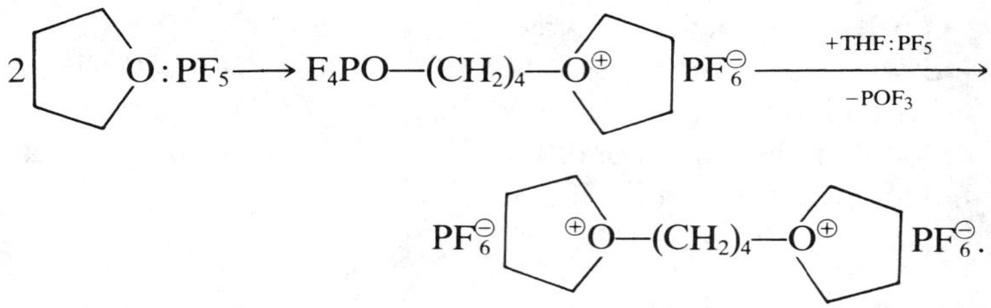

Earlier work with $(CH_3)_2O:PF_5$ lends support to this mechanism.[44] $(CH_3)_2O:PF_5$ was shown to rearrange at room temperature to give $(CH_3)_3O^+PF_6^-$ and POF_3. Formation of dication also agrees with work done by Sims, who proposed formation of dication when PF_5 and epichlorohydrin were used to form the tertiary oxonium ion.[45] Apparently, similar reactions can occur with higher valence state halides from groups VB, VIA and VIB, because dry $NbCl_5$, $TaCl_5$, WCl_6 and SeF_6 alone can initiate the polymerization of THF.[16,24] No detailed mechanistic studies have been made with these other Lewis acids.

Initiation by PF_5 has also been invoked as the reason why $LiPF_6$ ($LiF:PF_5$) alone will initiate the polymerization of THF, while $LiBF_4$ and $LiClO_4$ will not. Eckstein and Dreyfuss[8] explain that the equilibrium between $LiPF_6$ and LiF plus PF_5

$$LiMX_n \rightleftharpoons LiX + MX_{n-1}$$

is sufficiently to the right so that enough PF_5 is available to initiate the polymerization. Only very small amounts of PF_5 are required.[17–18] The corresponding equilibrium between $LiBF_4$ and LiF and BF_3 does not lead to polymerization, because 5–10 mol% BF_3 is necessary for initiation to occur.[46]

In the trivalent state, elements from group VA terminate the polymerization of THF by nucleophilic attack of the remaining pair of electrons on the position α to the oxonium ion and lead to stable salts. Thus, ammonia and amines give ammonium salts, while triphenylphosphine and other trialkylphosphines give phosphonium salts:

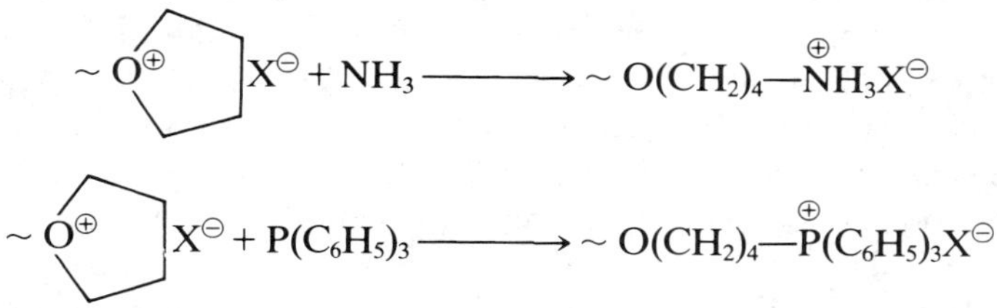

Group 2: In situ formation of tertiary oxonium ion from Lewis acid and reactive halide This process was discovered by Meerwein and coworkers in 1937.[7] The suggested reactions are as follows:

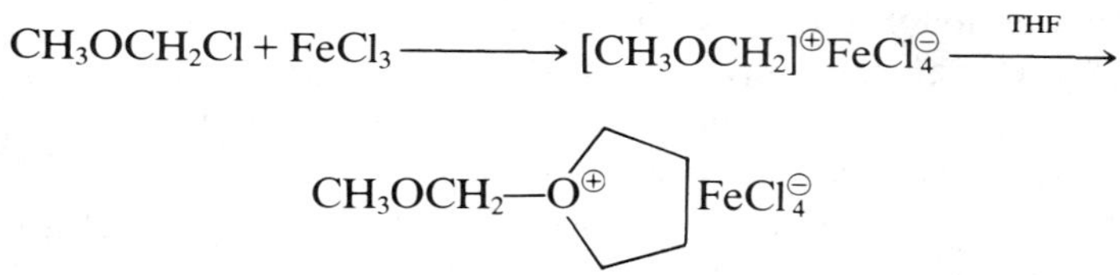

Benzyl chloride, benzal chloride, benzotrichloride and 1-chloromercapto-2,4-6-trichlorobenzene behave like chloromethylmethyl ether.

Group 2: In situ formation of tertiary oxonium ion from metal salt and reactive halides This process has been exploited relatively recently,[37,38,47–53] especially with alkyl halides, although the reaction with acyl halides has been known for a long time.[1] If, for example, acetyl chloride is treated with silver perchlorate in THF, polymerization ensues, presumably after intermediate formation of the acylium salt:

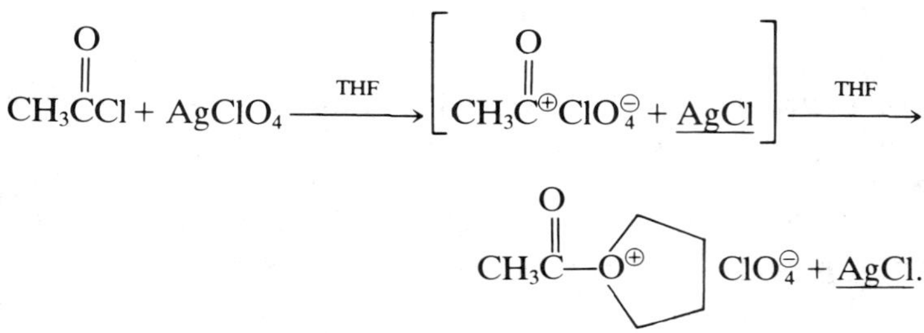

This reaction also occurs with most of the transition-metal perchlorates. The acylium salt can be isolated and purified first in a preliminary step and then the initiation proceeds according to the reaction shown in the discussion of group-1 mechanisms (p. 27). Formally, the same series of reactions can be written for alkyl halides and Franta *et al.*[38] present evidence that direct addition of the diphenylmethyl carbenium ion in $(C_6H_5)_2CH^{\oplus}SbF_6^{\ominus}$ to THF does occur and that $(C_6H_5)_2CH^{\oplus}SbF_6^{\ominus}$ can conveniently be formed *in situ* from $(C_6H_5)_2CHCl$ and $AgSbF_6$. Richards *et al.*[51] suggest a similar *in situ* mechanism with benzylchloride, but the preformed carbenium salt cannot be isolated. Dreyfuss *et al.*[54,55] were unable to find any evidence for intermediate formation of allyl carbenium ions in reactions between THF, allyl halide and silver salts, although they have evidence for the alkylation of THF. In particular, they have shown that pure graft copolymers are formed if the alkyl halide is a tertiary allylic halide, as can be found in polychloroprene, and they also find evidence for benzyl groups in the UV spectrum of polytetrahydrofurans initiated with benzyl chloride and silver salts. Because of the instability of the proposed carbenium-ion intermediates, they prefer to write what amounts to a silver-assisted displacement reaction:

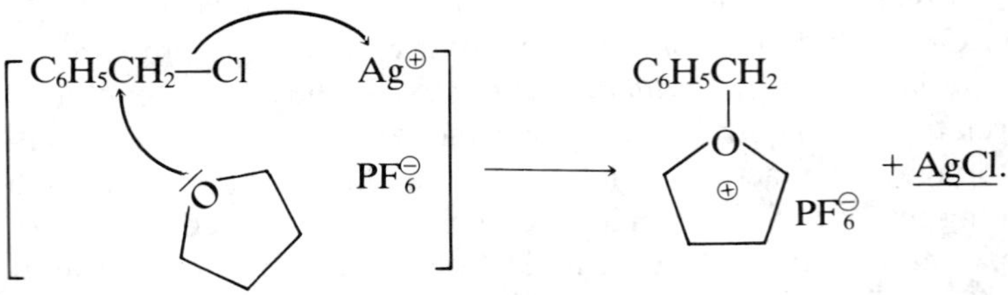

Other metals, especially the transition metals, behave much like silver.[8] In a study with 22 different perchlorate salts the order of reactivity based on percentage conversion after a given time and type of halide with which reaction would occur was Ag^+, $Hg^{2+} > Pb^{2+}$, $Tl^{3+} > Fe^{3+}$, Al^{3+}, Na^+, $Fe^{2+} > Cr^{3+}$, Mn^{2+}, Ni^{2+}, Cu^{2+}, Co^{2+}, $Cd^{2+} \gg Ce^{3+}$, Mg^{2+}, Ca^{2+}, Ba^{2+}, Zn^{2+}, Li^+, K^+, $NH_4^+ = 0$. Only Ag^+ and Hg^{2+} perchlorates gave PTHF when primary and secondary allyl chlorides were used. With the other perchlorates the order of reactivity of the halides was acetyl > benzoyl ≫ allyl.

With polychloroprene (Neoprene W), in which the tertiary allylic halides are the active sites, PTHF formed when using Ag^+, Hg^{2+}, Pb^{2+}, Tl^{3+}, Fe^{3+}, Al^{3+} and Na^+, but conversions decreased from 75% for Ag^+ and Hg^{2+} to 5% for Al^{3+}. Also, multiple peaks in the GPC tracings of polymers prepared using metals other than Ag^+ indicated that side reactions such as homopolymerization and termination were occurring; Ag^+ is the ion of choice for the initiation of THF polymerization by this method. The requirements for any reaction at all to occur are:

a) the metal salt must be soluble;

b) a stable oxonium-anion complex must form;

c) the products of the reaction must be more stable than the reactants.

Group 2: In situ formation of tertiary oxonium ion from $NOPF_6$ or NO_2PF_6 and halide This process was discovered by Eckstein and Dreyfuss.[13] The mechanism has not been studied, but it probably proceeds in the same way as the reaction above involving benzyl halide, THF and $AgPF_6$. The driving force for the reaction is formation of gaseous NO_2Cl or NOCl. Olah and coworkers[56] have studied the reaction between alkyl halides and NO_2PF_6 or $NOPF_6$ in acetonitrile. Following their suggestions, the reactions might also be written:

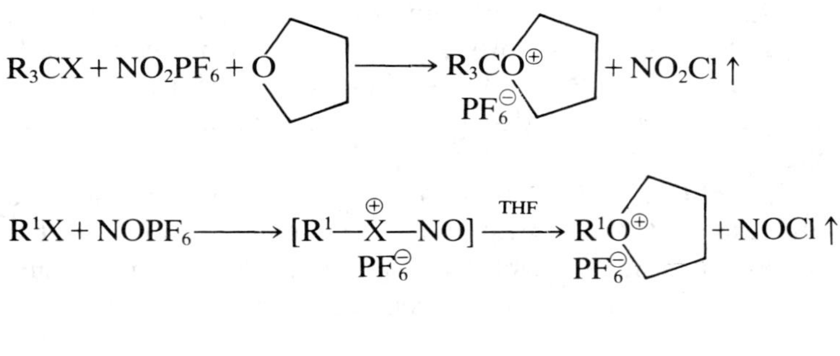

where R^1 = alkyl, aryl; X = F, Cl, Br, I. Experimentally, the reaction goes faster and more efficiently with $NOPF_6$.[13] In fact, with a very reactive halide, such as a tertiary allylic halide, initiation is nearly instantaneous.

Group 2: In situ formation of tertiary oxonium ion from other mixtures One example of this process is tris(trimethylsilyl)antimony with pentacarbonyl-(tetrahydrofuran)tungsten. Breunig and Wöhrle, who discovered this initiator,[57] propose a sequence of reactions leading to formation of polymer with a trimethylsilyl endgroup:

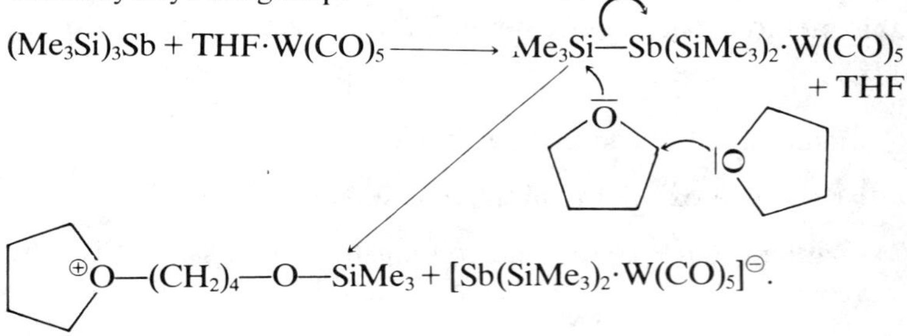

Group 3: Strong-protonic-acid addition to THF Such an addition process, with the formation of a secondary oxonium ion, can lead to the polymerization of THF:[1,34]

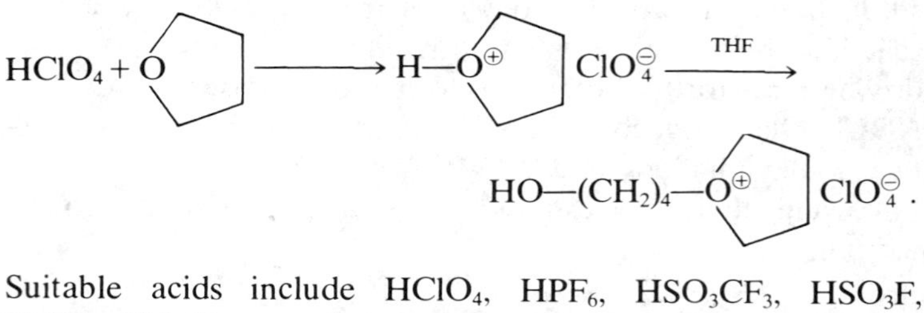

Suitable acids include $HClO_4$, HPF_6, HSO_3CF_3, HSO_3F, H_2SO_4, $HAsF_6$, etc. Conversion to polymer, the kinetics of polymerization and the properties of the polymer vary greatly with the anion. A few additional comments about counterions and their effects on acid stability are appropriate at this point. Within a given group of the periodic table, the order of stability of the anion seems to increase down the periodic table with respect to the central atom of the complex anion, but is greatest for the most electronegative of the elements surrounding the central atom. Thus, considering group VA, the order of stability and of reactivity is $HSbF_6 > HAsF_6 > HPF_6$ but, considering group VIIA, $HSbF_6 > HSbCl_6$. In addition, $HPF_6 > HBF_4 \gg HAlCl_4$. Again considering group VA, the larger the negatively charged ion, the more loosely it is bound to the proton and the faster initiation occurs.[15,58–61]

The cyclic secondary oxonium ion is a relatively unreactive species. Thus, initiation by protonic acid is a comparatively slow reaction. The presence of hydroxyl end groups also has significant effects on the molecular weight of the products formed.[62] These effects on molecular weight will be discussed in Chapter 5, which describes solution properties.

Group 3: In situ formation of protonic acid This process has been recognized comparatively recently as an important method of initiation of THF polymerization. Even now, as will become clear from the evidence presented here, not all investigators accept this mechanism. It is suggested, however, that the material summarized here is convincing. Because of the controversy and confusion that exist, this subject will be allotted a lengthier discussion than is really justified in comparison with the other mechanisms.

1) *In situ* formation of protonic acid from carbenium ions and THF was first proposed by Dreyfuss *et al.*[63] after they observed a far-downfield proton (~850 ppm relative to Me_4Si) in the ^1H NMR spectrum, in addition to the previously known triphenylmethane,[64–67] in reaction mixtures of 2-methyl-tetrahydrofuran or THF and triphenylmethyl hexachloroantimonate or hexafluorophosphate:

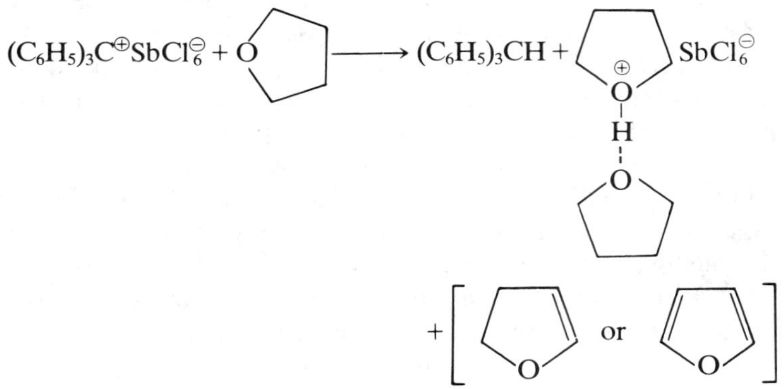

In this equation the proton is shown shared by two THF molecules. This is a more accurate description of the real situation than the secondary oxonium ion shown in the previous section (p. 34), where one THF molecule has been omitted for simplicity.[68]

In situ formation of protonic acid has now been confirmed in polymerizations initiated by quaternary ammonium salts plus electric current, by halonium, triarylsulfonium and triarylselenonium salts plus light, by mercuric perchlorate, $NOPF_6$ or NO_2PF_6, and may also be occurring with other reagents to be discussed below. The usual suggestion is that a hydrogen in some form (H^+, H^-, $H^\cdot$) is abstracted initially from a THF molecule. Strong protonic acid, which initiates polymerization as shown above, forms in subsequent reactions. The abstracted THF molecule decomposes to form a species such as furan or dihydrofuran, which is destroyed by the strong acid. When protonic acid forms *in situ*, polymerization reaction mixtures are highly colored, ranging from pink to purple to brown.

The formation of protonic acid from carbenium ions and THF is a more common reaction than the addition observed by Franta *et al.* for diphenylmethyl carbenium ion.[38] For example, $C_6H_5N_2^{\oplus}PF_6^{\ominus}$ behaves like triphenylmethyl salts, and benzene is formed as a side product. Attempts to form block copolymers by adding polymers terminated by carbenium ions to THF have led to mixtures of homopolymers.[1] Conversely, oxonium ions do not initiate the cationic polymerization of most common vinyl monomers such as styrene.[69] There is at least one exception to this general rule. Cyclopentadiene can copolymerize with THF.[70,71] Discussion of the ingenious methods that have been developed to bond vinyl polymers and polyether polymers is reserved for Chapter 8.

2) *In situ* formation of protonic acid from $NOPF_6$ or NO_2PF_6 and THF was also discovered by Eckstein and Dreyfuss.[13] The mechanism of this reaction has not been studied thoroughly, but considering the work of Olah and Ho with alkyl methyl ethers and nitrosyl or nitronium tetrafluoroborates, the following equations can be written:[72,73]

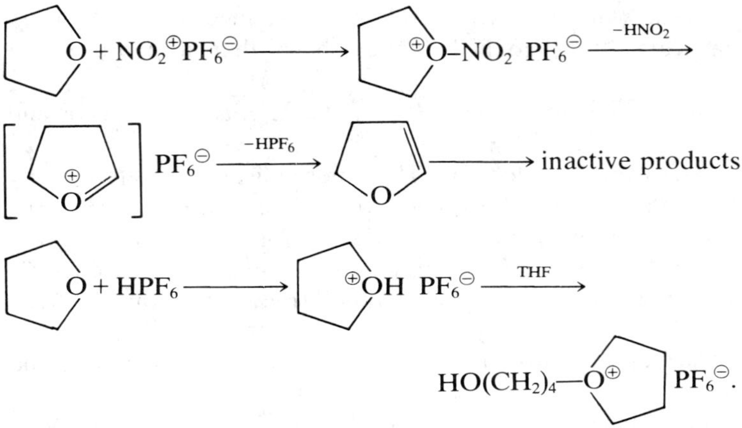

^{1}H NMR spectra of mixtures of THF and NO_2PF_6 do show a far-downfield proton.[13]

Similar reactions can be written for $NOPF_6$. As Olah *et al.* point out, the difference between the NO^+ and NO_2^+ ions is that NO^+ has the ability to abstract a hydride ion, whereas NO_2^+ is an oxidizing agent. Thus, an alternative mechanism to the above with $NOPF_6$ would omit the addition and would yield HPF_6 and unstable HNO directly.

3) Electrolytic initiation of THF polymerization has been observed by several groups.[9–12] Soluble quaternary ammonium salts such as $(C_4H_9)_4N^{\oplus}ClO_4^{\ominus}$ or $(C_2H_5)_4N^{\oplus}ClO_4^{\ominus}$ have been present when THF polymerization ensued, but the passage of an electric current through the solution was necessary.[9–11] Yamazaki and coworkers[9] studied these systems and concluded that initiation is independent of the cation and most probably results from interaction of the perchlorate radical and a proton:

$$ClO_4^{\ominus} + e^{\ominus} \longrightarrow ClO_4^{\bullet}$$

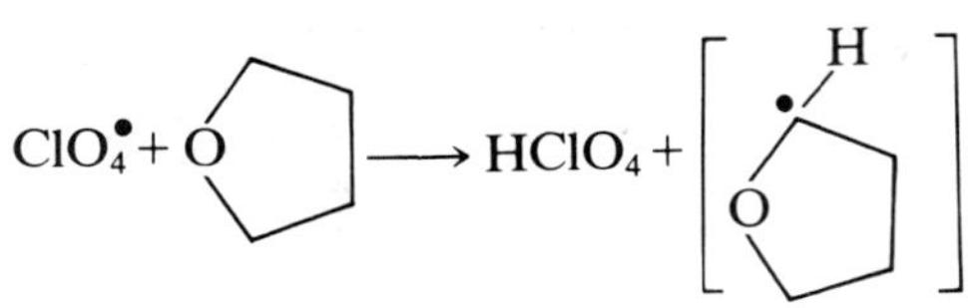

These workers have considered other mechanisms also.[9] In particular, they looked for evidence that a hydrogen-abstracted THF molecule,  $ClO_4^{\ominus}$, added to THF and initiated polymerization. This mechanism is preferred by Ledwith and coworkers whenever abstraction of THF occurs, although they have no evidence for the proposed addition.[64,66,74] In their studies of the mechanism of electrolytic initiation, Yamazaki and coworkers[9] tried to trap the aldehyde or acetal that would result if the  cation could initiate the polymerization of THF. Only 0–5% of the polytetrahydrofuran molecules were determined to have aldehyde groups, and even these could have arisen by some other mechanism such as oxidation of an –OH end-group. In contrast, hydroxyl-group determination in polytetrahydrofurans, which were electrolytically obtained and terminated by water, indicated that the hydroxyl-group content was in quantitative agreement with the assumption that the polymer had two hydroxyl end groups. This result is expected if protonic acid induces polymerization:

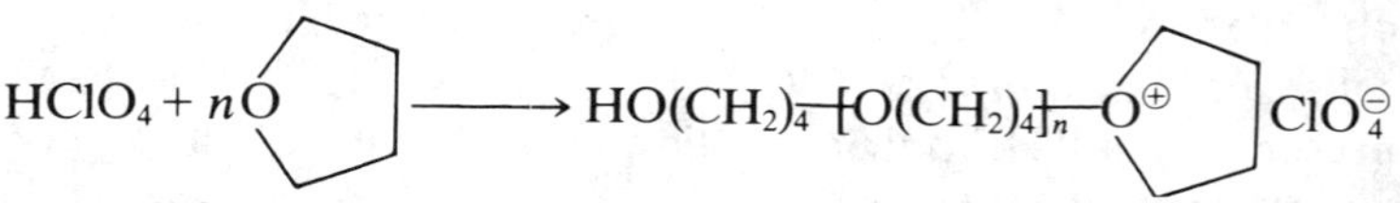

$$\xrightarrow{H_2O} HO(CH_2)_4[-O(CH_2)_4-]_nO(CH_2)_4-OH + HClO_4.$$

Further support for the instability of the intermediate cation, , is found in the work of Ward and Sherman,[75] who found evidence for the formation of 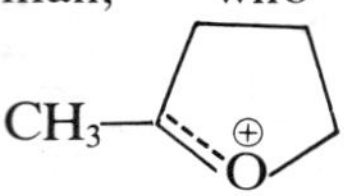 upon trifluoroacetolysis of 5-(*p*-bromobenzenesulfonoxy)-2-pentanone. However, in contrast to the phenyl analog, which was stable when formed under similar conditions, the oxonium ion with the methyl substituent decomposed quite rapidly to give the trifluoroacetate.[75] Their results suggest that the order of stability for the three ions under consideration would be:

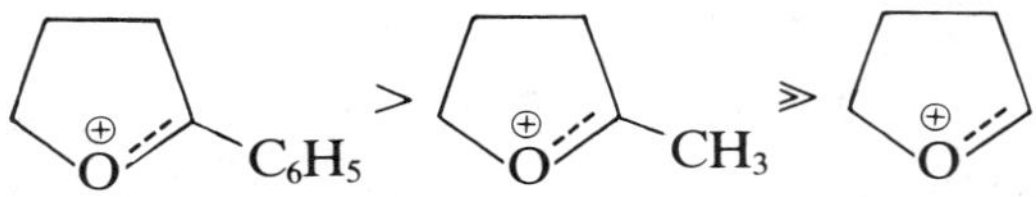

since the 2-carbon is a secondary ion in the latter, whereas in the others it is tertiary.

Electrolytic polymerization of THF also occurs in the presence of $LiClO_4$, most likely by the same mechanism. Note that $LiClO_4$ does not initiate THF polymerization alone or in the presence of halide.[8] Also NH_4ClO_4 is insoluble in THF and polymerization does not occur.[8]

4) Photoinitiation of THF polymerization occurs with a variety of materials that are photolytically unstable: $Ar_3S^{\oplus}$, $R_2X^{\oplus}$, $Ar_3Se^{\oplus}$. This work is relatively new and has provided the most convincing evidence to date that the *in situ* generation of protonic acid for the initiation of THF polymerization is more common than is generally accepted or appreciated.

Halonium ions from group VIIA do not initiate the polymerization of THF directly. In the absence of light no reaction occurs. On irradiation, photochemical reactions occur, protonic acid is generated and formation of tertiary oxonium ion occurs as shown, above (p. 34).[19,20] For example, diphenylbromonium hexafluorophosphate led to bromobenzene, benzene, biphenyl and HPF_6. All compounds except the protonic acid were separated and identified by combined gas–liquid chromatography and mass spectroscopy. The suggested mechanism is shown in the following series of equations:

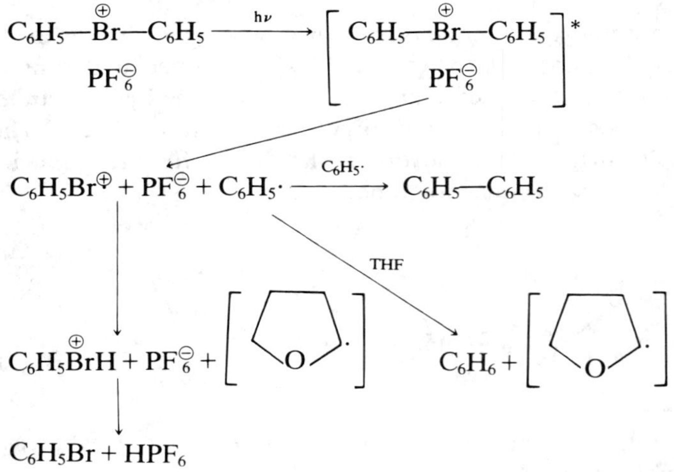

Triarylsulfonium salts and triarylselenonium salts undergo similar reactions.[15,35] Addition of certain dyes such as perylene allows the photolysis of these salts to proceed using visible light.[76] Further confirmation of the proposed mechanism was obtained by carrying out photolysis in acetone-d_6 and by carrying out crossover experiments using equivalent amounts of triphenylsulfonium and tris(4-methoxyphenyl)sulfonium hexafluoroarsenates. All the products consistent with the proposed mechanism were obtained.[15] In other experiments with the triarylsulfonium salts, mechanisms involving either participation of the monomer in a charge-transfer complex with the photoinitiator or initiation by attack of the nucleophilic monomer on the photoexcited salts were eliminated.[15] The results and the products were consistent only with initiation by the Brönsted acid corresponding to the counterion. Since no decomposition products of the acid were found, these investigators concluded that the counterion survives the photolysis intact.

Metal salts with stable counterions can lead to the formation of tertiary oxonium ions even when no reactive halide is present. Abdul-Rasoul *et al.*, have reported that the polymerization of THF is readily induced in the presence of $AgPF_6$

by thermal and photochemically active sources of free radicals.[74] They suggest that the important species must be obtained by hydrogen abstraction from THF and that the role of the $AgPF_6$ is to function as a one-electron oxidant for electron-donor radicals:

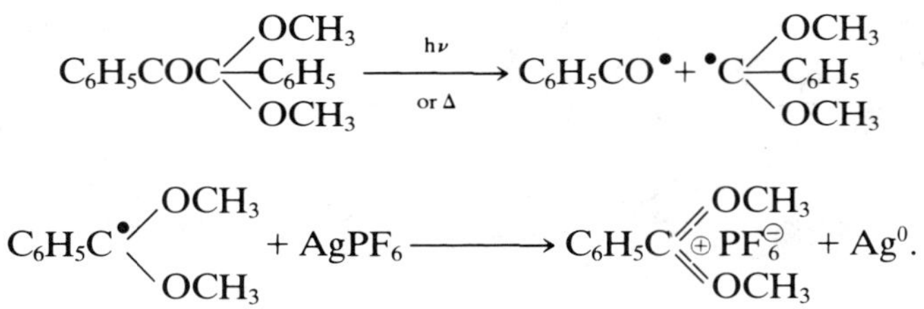

The carboxonium salt then probably initiates THF polymerization via direct alkylation by "exchange."

Similar reactions can occur after thermal or photochemical decomposition of phenylazotriphenylmethane to give triphenylmethyl radical. This radical also reacts with $AgPF_6$ in a one-electron oxidation reaction to give trityl cation, but then initiation proceeds via protonic acid

$$C_6H_5N{=}NC(C_6H_5)_3 \xrightarrow[\text{or } \Delta]{h\nu} C_6H_5 + N_2 + (C_6H_5)_3C^{\cdot}$$

$$(C_6H_5)_3C^{\cdot} + AgPF_6 \longrightarrow (C_6H_5)_3C^{\oplus}PF_6^{\ominus} + Ag^0.$$

Addition of other chemicals which serve as sources of radicals may not be necessary. Woodhouse et al. report that solutions of $AgBF_4$, $AgPF_6$, $AgSbF_6$, $AgOSO_2CF_3$, $(CuOSO_2CF_3)_2{\cdot}C_6H_6$ and $Cu(OSO_2CF_3)_2$ in anhydrous THF are stable indefinitely in the dark; but upon irradiation with 254-nm light the solutions become viscous, and elemental metal, Ag^0 or Cu^0, is formed.[77] Woodhouse *et al.* favor an initiation mechanism in which a metal-coordinated THF is activated for nucleophilic attack by a ligand-to-metal charge-transfer excitation.[77] In view of the total information now available, it seems preferable to take, as the major reaction path, a mechanism somewhat analogous to the equations above:

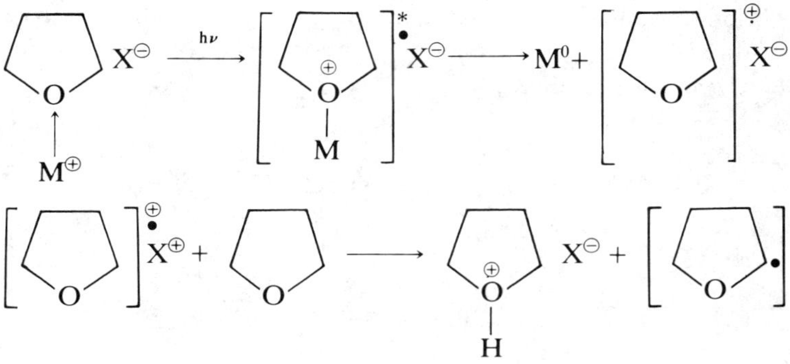

This mechanism is in agreement with a suggestion made by Yoshida and Hayashi, who found that irradiated 2-methyltetrahydrofuran glass gives an ESR signal. They explained their observation in the following way[78]:

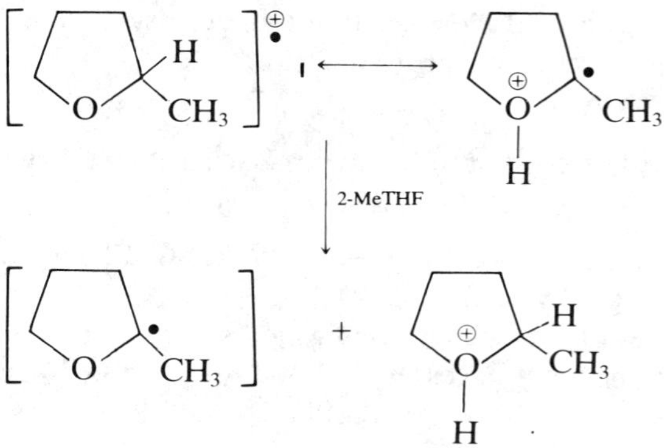

It is also in agreement with the ^1H NMR results for $Hg(ClO_4)_2$ in THF. In the absence of halide, when $Hg(ClO_4)_2$ is added to THF, protonic acid and a black precipitate, most likely mercury, form.[8]

5) Vinylidene cyanide reacts with THF in toluene at 25°C to give the homopolymer of both monomers.[79] Oguni *et al.*,

suggest that the vinylidene cyanide reacts with THF to form an intermediate donor–acceptor complex, which can undergo a one-electron transfer followed by hydrogen abstraction and cationic polymerization of the THF and anionic polymerization of the vinylidene cyanide:

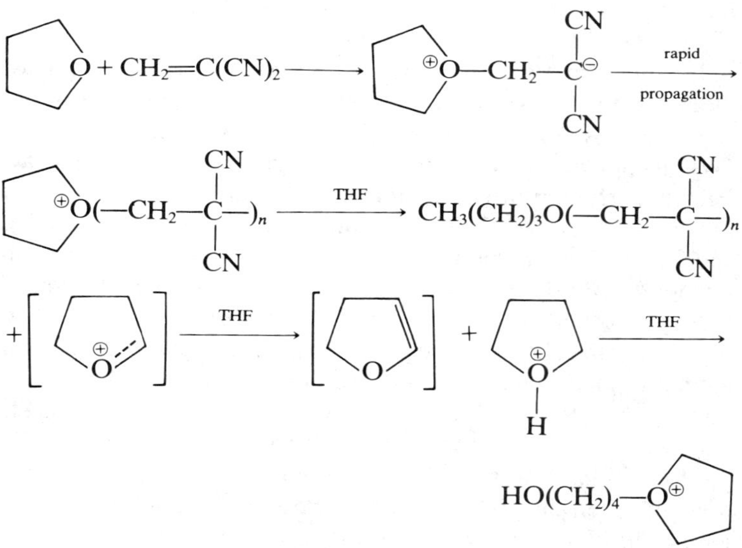

Summary of relative reactivities The relative reactivity of compounds for inducing the polymerization of THF can be related to the oxidation state of the principal elements in the compounds and to the position of those elements in the periodic table. Except for group VIIA, and excluding the inert gases, which are unreactive for present considerations, compounds whose elements are on the right of the periodic table are more reactive than those whose elements are on the left; reactivity decreases across the periodic table from right to left. There are no reports that halonium ions from group VIIA add to THF and lead to tertiary oxonium ions. Elements from group VIIA are useful mainly in counterions or when associated with other principal elements as in PF_5. When halonium ions are used, protonic acid is formed by a photolytic reaction and the polymerization of THF is induced by protonic acid. Hydrogen is in a class by itself and its activity is related to the anion associated with it. Direct addition to form tertiary oxonium ions occurs only with acyl ions and perhaps with $(C_5H_5)_2CH^{\oplus}$. The dioxolenium ion, which initiates via direct alkylation by "exchange", also results in addition, but addition is an adventitious result of its cyclic structure.

In situ formation of tertiary oxonium ions from Lewis acids alone occurs with compounds whose principal elements are from groups VA (PF_5), VB ($TaCl_5$), VIA (SeF_6) and VIB (WCl_6); not all elements from these groups react in this way. Elements from group III either in Lewis acids—BF_3, $AlEt_3$—or in ions—Al^{3+} in $Al(ClO_4)_3$—require the addition of a reactive species such as epoxide or halide, respectively. The transition elements are useful as cations, and they require the addition of halide or other activation such as light. For example, Ag^+ in $AgClO_4$, can initiate polymerization photolytically. Salts from elements in groups IIA and IIIB are completely unreactive. Salts from elements in group IA are reactive if their salts dissolve in THF, but their reactivity varies greatly. For example $LiClO_4$ leads to polymerization only if an electric current is flowing. Even then it is the perchlorate rather than the Li^+ that is responsible for reactions leading to initiation; the Li^+ is only a means of introducing the perchlorate. $LiPF_6$ and $NaClO_4$ lead to polymerization in the presence of reactive halide. While $LiPF_6$ can be used alone, it seems most likely that PF_5 is the reagent responsible for the initiation of THF polymerization.

This analysis of initiation based on the periodic table combined with chemical structure should allow the usefulness of most simple substances for initiating THF polymerization to be predicted. Suitable examples are $GeCl_4$, $GeCl_2$ and Sr compounds, for which there are no literature data. Noting that Ge is in group IVA, it may be supposed that $GeCl_4$ can probably be used to induce THF polymerization, but addition of epoxide or reactive halide will probably be necessary, while $GeCl_2$ probably will be unreactive under the same conditions. Compounds from

Sr, which is in group IIA, will probably be unreactive even with additives.

3.2.2 Propagation–Depropagation

THF polymerizes only by a ring-opening mechanism, and an equilibrium between monomer and polymer is reached at all temperatures and concentrations. In general, propagation occurs by an S_N2 mechanism as a result of bimolecular collisions between THF monomer and the tertiary-oxonium-ion chain end. Only collisions at the α-carbon atoms of the oxonium ion result in chain growth. Depropagation occurs by an intramolecular nucleophilic attack of the penultimate oxygen atom of the polymer chain followed by expulsion of monomer[1–5]:

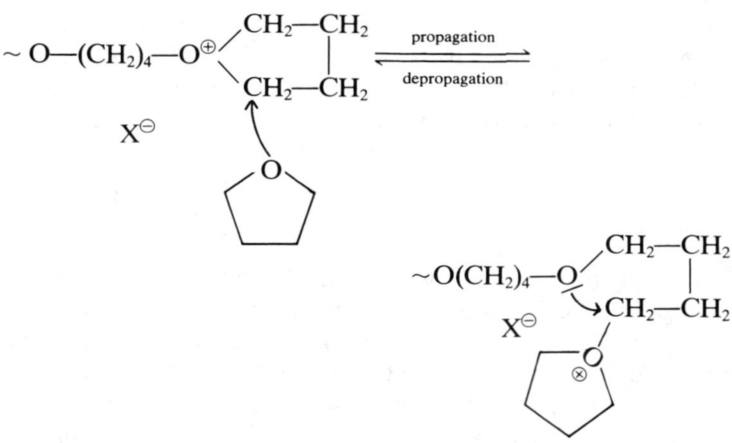

Historically,[2] the possibility that carbenium ions (carbocations) are involved in the propagation was suggested, but the accumulated evidence is overwhelmingly in support of chain growth through oxonium-ion intermediates.[1,5]

The question of whether free ions or ion pairs are involved in chain growth has been studied.[80,81] Conductimetric and kinetic data in methylene chloride indicate that the reactivity of the free oxonium ion is only seven times greater than the reactivity of the ion pair from THF oxonium ion and its counterion, and so growth of both species must occur. This contrasts sharply with anionic polymerizations, where free ions propagate 1,000 times faster than ion pairs.[82]

With a few counterions, an equilibrium between the THF oxonium ion and an ester of the counterion can occur[32,83–90]

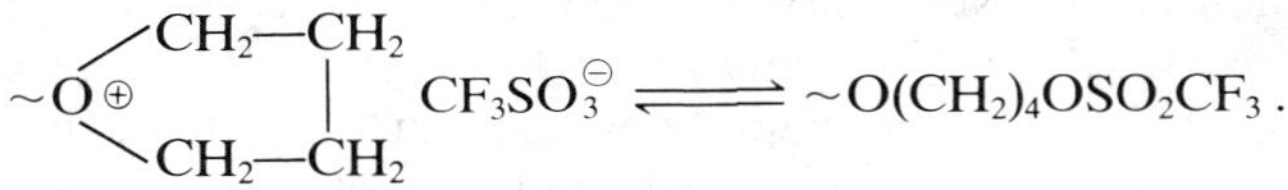

The relative concentration of each species depends upon the dielectric constant of the medium. Both species can propagate but, as will be described in Chapter 4, the ion is very much more reactive. Only counterions from super acids, i.e., acids that are stronger than 100% sulfuric acid based on their ability to donate a proton to a weak base and on the extent of autoprotolysis,[91,92] behave in this way. Examples include the trifluoromethanesulfonic acid illustrated above as well as fluorosulfonic, chlorosulfonic and 2,4,6-trinitrobenzenesulfonic acids.[36]

3.2.3 Termination

THF polymerizations can be terminated by a variety of nucleophilic reagents, including common materials like water and oxygen. If a "living" polymerization is desired, care must be taken to eliminate adventitious impurities. Unwanted termination reactions with counterion leading to uncontrolled end groups can also occur. Hence, as already pointed out, the choice of counterion is very important. Still, it is possible to conduct a THF polymerization in such a way that almost any desired end group can be introduced. Industrial applications of PTHF are based largely on the ability to prepare α,ω-dihydroxy PTHFs and, so this ability to control termination reactions is very important. In addition, many theoretical studies of kinetics and structure rely on the ability to control termination. In all cases, termination leads to ring opening of the oxonium-ion end group and formation of a stable product, which is useful for further reaction or determination. Only a brief review of termination possibilities will be given here. Discussion of methods by which dihydroxy PTHF of controlled molecular weight may be prepared commercially are reserved for Chapter 7.

With common nucleophiles The oxonium-ion end group can react with water, alcohols, ammonia, amines, carboxylic acids, etc., to give a wide variety of end groups including alcohols, ethers, amines, esters, etc.:

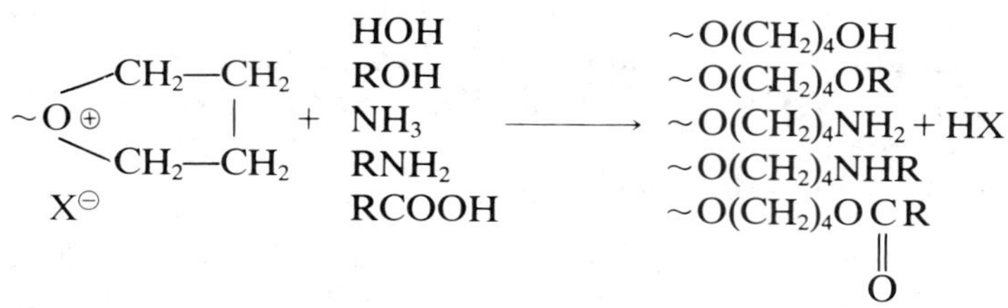

For end-group analysis The main features of THF polymerization are so well established that the polymerization of THF has become a model for the study of oxonium-ion polymerization. Thus, there has been a need both in kinetic analyses and in the study of graft copolymers to determine the number of active centers present at any given time. Only a small number of reagents have proven to be useful for end-group analysis and only one of them was developed long enough ago to have been used very much. Low-molecular-weight polymers, 1,000–2,000 g $mole^{-1}$, can have end groups determined by classical methods such as hydroxyl-number determination[93] or halide determination, and will not be discussed here. To cover a broad spectrum of molecular weights, more sensitive techniques are required. The following examples are typical of techniques that can be used.

The phenoxy end-capping method This, the oldest and most widely applied method,[4,94] is based on the quantitative conversion of the propagating ion into the corresponding phenyl ether, by treatment with sodium phenoxide and analysis for the phenyl ether group by UV spectroscopy:

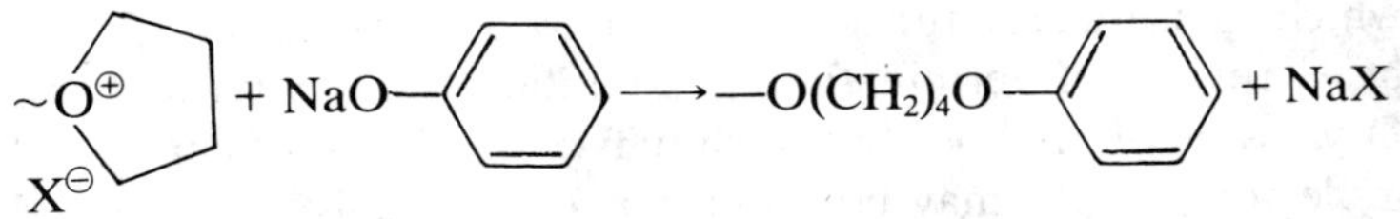

For those who may wish to apply the method, the directions given by the authors are included here:[11]

> At a desired time in the polymerization reaction, a solution of sodium phenoxide in THF is added to terminate the polymerization. The short-stopping mixture is stirred for one hour at room temperature, washed with a large excess of 1N aqueous NaOH solution and then extracted with CH_2Cl_2. The extract is dried over anhydrous K_2CO_3, centrifuged and subjected to UV measurement.

With initiators such as trialkyloxonium salts, phenetole is formed as a side product, and must be separated from the polymer prior to the UV measurement. This can be accomplished by vacuum distillation with the aid of decalin as a distillation entrainer (0.2–0.3 mm Hg at room temperature).

In methylene chloride and in diethyl ether, the phenyl-ether end group has λ_{max} at 272 nm and ϵ_{max} of 1.93×10^3 l mol^{-1} cm^{-1}.[4]

This method has been applied successfully to polymerizations with counterions including BF_4^- and related compounds, the counterion formed in $AlEt_3$–H_2O-epichlorohydrin-, $SnCl_4$-epichlorohydrin- and $EtAlCl_2$-epichlorohydrin-initiated polymerizations, $AlCl_4^-$ and superacid anions.[4] It has been used less successfully with polymerizations containing PF_6^- and SbF_6^- anions, and the method is not recommended with these anions.[95] It fails with PF_5-initiated polymerizations,[96,97] because the residual PF_5 reacts quantitatively with the phenoxide.[97] The reaction products formed cannot be removed by extraction with NaOH solution, and absorb in the same part of the UV as the phenoxy group on the polymer.[97] It cannot be used if the system contains other chromaphores that absorb in the 272-nm region, which is true in the case of many of the block and graft copolymers containing PTHF (see Chapters 7 and 8.)

Triphenylphosphine Triphenylphosphine has been used in combination with UV analysis in much the same manner as sodium phenoxide;[96] see p. 31 for the reaction. This reagent has some advantages over sodium phenoxide if the initiating system contains silver salts, and it can be used if PF_6^- or SbF_6^- counterions are present. Its principal disadvantage is that the phosphonium ions formed are sensitive to oxidation to phosphine oxides, so that determination of end-group concentration should be carried out as soon as possible after termination.[98,99] The method cannot be used if the system contains a chromaphore that absorbs in the same region as the phosphonium ion.

NMR If the peaks due to initiator, polymer, oxonium ion and side products, if any, are sufficiently separated in the NMR spectra of polymerizing mixtures, the polymerization reaction need not be terminated. NMR is a powerful tool for studying such reactions. ^{1}H NMR (especially at 100–300 MHz), ^{13}C NMR and ^{19}F NMR have been very useful for studying polymerizations containing anions from superacids and the esters that are formed from them.[86–90] However, the relatively large concentration of initiator and active ends that NMR analysis often requires can sometimes lead to a misinterpretation of the data, and so care should be used in such analyses.

Several investigators have chosen to terminate THF polymerizations with reagents that can later be used for analysis by NMR. Suitable reagents include ^{13}C labelled sodium methylate,[5] triethylammonium picrate[5] and triphenylphosphine.[100,101] In the latter case ^{31}P NMR was used; the method works quite satisfactorily even if PF_6^- or SbF_6^- anions or a UV-absorbing chromaphore is present, but it has the same disadvantages with regard to oxidation sensitivity described at the top of this page. If precipitates are present, as they will be when silver salts are part of the initiating system, removal of the salts is necessary before ^{31}P NMR spectroscopy can be carried out.

Termination with ammonia Termination with ammonia or ammonium-hydroxide–ammonium-chloride buffer solution, followed by determination of the resulting primary amine concentration using fluorescamine *and fluorescence spectroscopy*, is a very recent development.[102]

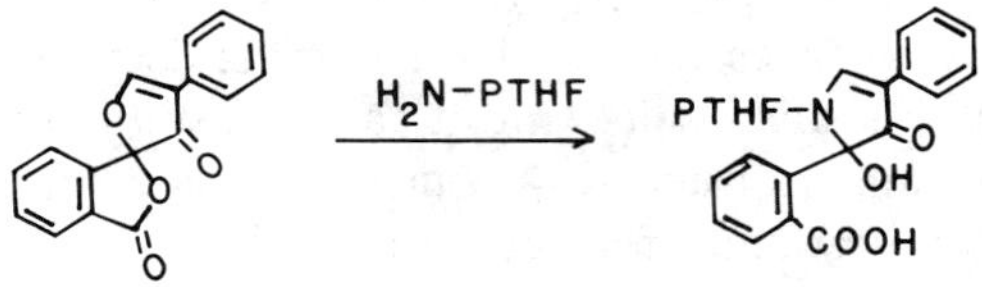

The method is sensitive even at concentrations of 10^{-6} M and below. It can be used with SbF_6^- and PF_6^- counterions. The absorbances of the chromaphores of backbones used to prepare graft copolymers and of silver salts do not interfere. When Eckstein and Dreyfuss[102] compared this new method to the "phenoxy end-capping method" and to termination by triphenylphosphine followed by ^{31}P NMR spectroscopy, they found that, especially for graft copolymers, it was easier and more reliable than either of the other methods.[101] The procedure is as follows. At the selected time in the polymerization reaction, a large excess of NH_4OH–NH_4Cl buffer solution (0.087 g of NH_4Cl per ml of concentrated NH_4OH) is added to terminate the polymerization. (Alternatively, a large excess of gaseous ammonia in THF may be added.) The mixture is stirred or shaken for 30 min at room temperature, the liquid is poured off and the precipitated polymer is washed with methanol to remove NH_4Cl. (For low-molecular-weight polymers, <2000 g/mol^{-1}, the buffer-solution mixture is poured onto ice, and the polymer can be extracted with chloroform and dried in the same way as in the "phenoxy end-capping" method.) The polymer is dried in a vacuum oven, and can be stored safely until ready to use, since the amine groups formed are stable. Between 20 and 50 mg are dissolved in 50 ml chloroform in a 50 ml volumetric flask. (The larger weights are used for high-molecular-weight polymers and graft copolymers.) A known volume, 10–50 μl, of the polymer solution, 200 μl of fluoroescamine stock solution (2×10^{-3} M) and 400 μl of 95% ethanol are added to a 10 ml volumetric flask and thoroughly mixed after each addition. After diluting with $CHCl_3$ to volume, the emission spectrum of the reaction product is measured at 485 nm by excitation to 395 nm. Either 1,4-

diaminobutane or amine terminated polytetrahydrofuran with known amine content can be used as standard. The similarity of their excitation spectra monitored at 485 nm emission is shown in Figure 3.4.

The principal disadvantage of this method is that it is so new it has not been tested sufficiently. Certain difficulties may already be anticipated, however. Some investigators may have difficulty in obtaining quantitative conversion of the oxonium ions to primary amine groups. Unless this is done properly,

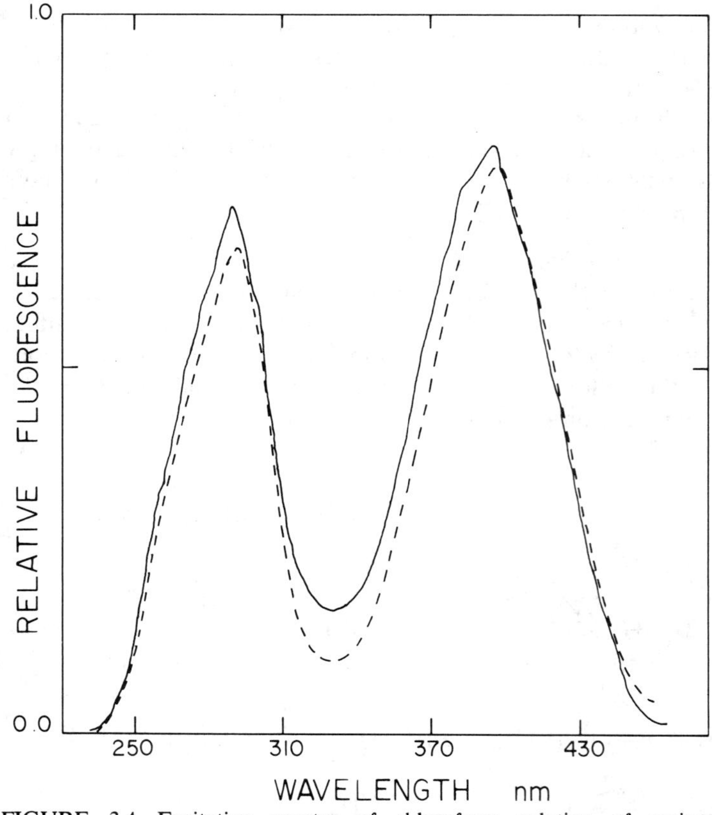

FIGURE 3.4 Excitation spectra of chloroform solution of amines + fluorescamine + 95% ethanol monitored at 485 nm. Emission corresponding to 1,4-diaminobutane- (continuous curve) and amine-terminated PTHF (dashed curve) is shown.

some secondary amines may form[103] and the resulting amine-determination result will be low. Fluorescamine does not react with secondary amines. If the alternative method of termination with NH_4OH–NH_4Cl buffer solution is used, corrections are necessary. According to the report 75% amine groups, 15% hydroxyl groups, 10% chlorine groups, and apparently no secondary amines are formed. While it is claimed that this number is reproducible, and the results do agree with those from ^{31}P NMR spectroscopy,[101] some investigators may not feel comfortable with a mixture.

With anionic "living" polymer Oxonium-ion end groups react with the "living" anions present in polymers prepared by anionic polymerization to give block copolymers.[104–106] Although the reaction has been known since 1966, and can proceed cleanly to yield pure block copolymers, surprisingly few such polymers have been characterized. Those that have will be discussed in Chapter 8.

With counterion Reaction with counterion results in a decrease in the number of active centers. At polymerization temperatures near 0°C choice of counterion is not so limited. However, by 25°C with BF_4^-, transfer and elimination reactions like the following begin to have a substantial effect on the polymerization kinetics:[5,31,107,108]

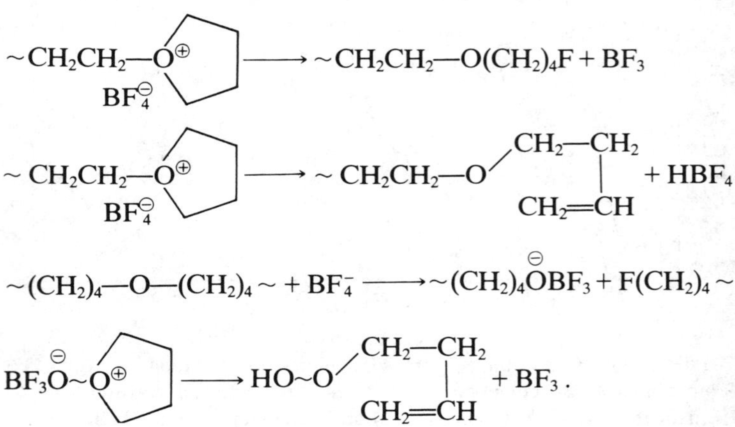

With boron-containing counterions, these reactions constitute true termination, because BF_3 alone can initiate polymerization of THF only at very high concentrations.[46]

Termination reactions are also important with $AlCl_4^-$ counterions even at 0°C.[109] Counterions SbF_6^- and PF_6^- are reasonably stable for moderate times up to the ceiling temperature for THF polymerization but, after long times at high temperatures, some reaction will also occur. The $SbCl_6^-$ counterion shows intermediate behavior. It will react with the oxonium ion to terminate a chain and form $SbCl_5$. Like other Group VA Lewis acids, $SbCl_5$ can initiate the polymerization of THF. See the discussion in Section 3.2.1 (p. 29). The reaction with $SbCl_6^-$ counterion must thus be classified as combined termination and transfer.

The counterions from nonhydrolyzable superacids are quite stable under polymerization conditions, since fragmentation of the noncomplex anions is not possible.[110] However, as already noted, reaction to form an ester can occur. Since the esters propagate at a much slower rate than the ions (see Chapter 4), pseudotermination is inferred. The amount of ester formed is very strongly solvent-dependent.

Hydrolyzable protonic acids, such as chlorosulfonic acid, can react to form mono- and dialkyl sulfates:[2]

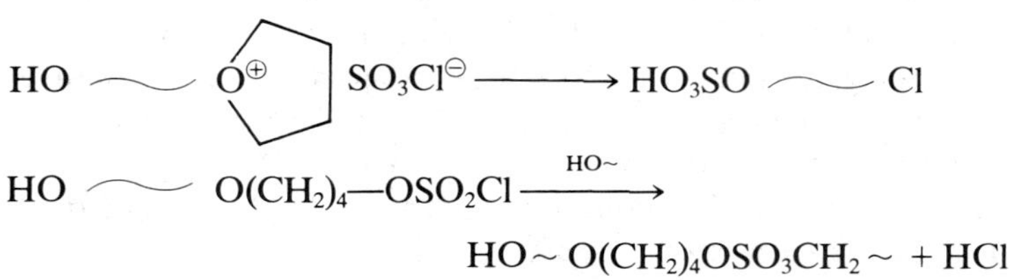

Sulfate formation is normally irreversible and is therefore a termination reaction.[111]

3.2.4 Transfer

When the polymerization of THF is initiated rapidly and efficiently, and a reasonably stable counterion is used, the polymerization belongs to the class called "living". The number-average molecular weight increases in direct proportion to the degree of conversion, and the concentration of growing centers is equal or close to the initial initiator concentration. If more monomer is added after equilibrium conversion has been attained, polymerization will resume again at close to the original rate. Polymers with low polydispersity can be prepared.

In these respects, the cationic polymerization of THF closely resembles the anionic polymerization of styrene. It differs in that, in the anionic polymerization of styrene, polymers with a narrow molecular-weight distribution are readily formed at high conversion whereas, in THF polymerizations, PTHF with a narrow molecular-weight distribution can be obtained routinely only at low conversions and short times, or if the propagation proceeds at low temperature. Since at these low temperatures the rate of initiation is usually not instantaneous, some special conditions, described in Section 5.6. have to be used.

One reason for this difference between molecular-weight distributions formed in THF and in anionic polymerizations lies in differences in the rates of depropagation for the two systems. "Living" PTHF depropagates at a significant rate at the temperatures where propagation occurs at a reasonable rate. Heating a "living" PTHF at equilibrium conversion to the ceiling temperature gives pure monomer in a short time. Heating a "living" polystyrene leads to a very small change in monomer concentration. Miyake and Stockmayer have developed theoretical equations that permit estimates of these times to be calculated.[112] Based on the rates of depropagation alone, the time required for the molecular-weight distributions to become Gaussian will be 20 days for PTHF and 80 years for polystyrene.[58]

A more important reason lies in intermolecular transfer reactions that can occur with PTHF but not with "living" anionic polymers. PTHF has oxygen atoms regularly spaced along its backbone, and these oxygen atoms can react with the THF oxonium ion to produce a branched ion:

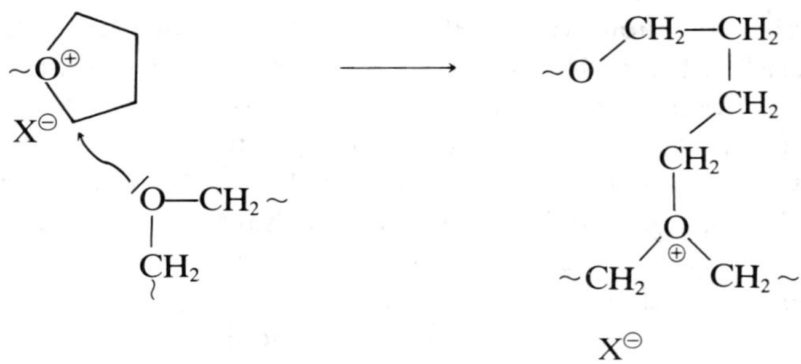

Further reaction with THF can occur at any of the three carbon atoms α to the acyclic polymeric oxonium ion. The result is a randomization of the molecular weight of PTHF within eight hours, under conditions where 20 days would be predicted on the basis of depropagation alone.

Other oxygen-containing materials such as ethers,[26] water,[113] alcohols and especially orthoformates[26] undergo similar reactions with the THF oxonium ion. These transfer reactions result in chain scission and reduce the molecular weight of the PTHF. Possible transfer reactions with a counterion like $SbCl_6^-$ have already been alluded to (pp. 52–53).

As already noted, depropagation with expulsion of monomer occurs by intramolecular reaction of the penultimate oxygen and the acyclic carbon α to the oxonium ion. Another product that can be postulated as a result of intramolecular reaction with polymer oxygen is a macrocyclic oxonium ion:[114]

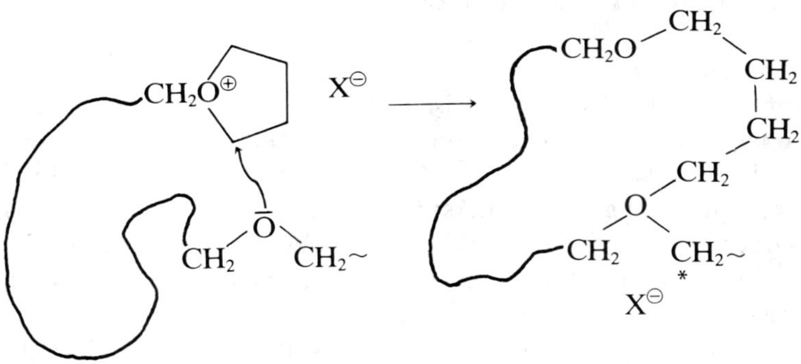

Further reaction with THF at the carbon labelled with an asterisk will lead to the formation of macrocycles. With some cyclic compounds such as ethylene oxide, oxetane and thietanes, a major portion of the product of ring-opening reactions is composed of macrocycles.[115–119] In THF polymerizations the number of macrocycles formed is so small that, for many years, chemists thought that none forms.[1]

In 1977 and 1978 McKenna *et al.*[120,121] used a combination of gas chromatography and chemical ionization mass spectrometry to identify a series of linear and macrocyclic oligomers formed during polymerizations of THF initiated by $Me_3O^+BF_4^-$, $CF_3SO_3CH_3$, $CF_3SO_3C_2H_5$ or CF_3SO_3H in nitromethane. The proportions of dimer, trimer, tetramer, . . . , octomer depended on initiator and polymerization conditions. The concentration of the macrocycles was generally less than 3% of the total product formed at 25°C, much smaller than with, say, ethylene oxide or oxetane, where the cyclic-oligomer concentration is in the range 30–100%. Both cyclic and acyclic oligomers formed with the trimethyloxonium salt and with the esters; only cyclic oligomers formed when the acid was used. The molecular weight, rate of initiation and number of active centers were all very different at comparable initiator concentrations and polymerization times after initiation by free acid. These effects will be considered in greater detail in Chapter 5.

3.3 SUBSTITUTED TETRAHYDROFURANS

The polymerization of monosubstituted THFs is not expected on a thermodynamic basis[27,122–4] and high-molecular-weight polymers have not been prepared from this type of monomer. Polymerization of the 3-methyl derivative has yielded low-molecular-weight polymers;[125] the 2-methyl derivative was converted to small oligomers with a degree of polymerization of about three;[87] but 3-chloromethyltetrahydrofuran has so far resisted all attempts at polymerization.[126]

As far as it has been possible to establish, except for the

special case of bicyclic monomers, the polymerization of THFs with more than one substituent has not been studied. Bicyclic structures in which one of the rings is a THF molecule are strained compounds with H–H crowding. Many such monomers are known to polymerize. Both ring strain and H–H crowding are relieved on polymerization. Table 3.1 gives the polymerizability of bicyclic THFs that have been studied. Most of them polymerize readily. However, the mode of fusion in bond-bridged bicyclic ethers has a large effect on polymerizability. The studies of Kops and coworkers indicate that *trans* isomers can readily be converted to polymers with high molecular weight, whereas the *cis* isomers are not homopolymerizable (see table 3.1). Only *trans*-7-oxabicyclo[4·3·0]nonane is capable of reversible polymerization.[128] With the other monomers, depolymerization is absent.[145–6] Studies of the mechanism of polymerization of *exo* and *endo* isomers indicate that ring opening proceeds exclusively through nucleophilic attack on the activated ring structure at C-4.[144–5,147] Both the *exo* and *endo* derivatives lead to polymers in which succeeding ether groups are *trans* to each other.[145]

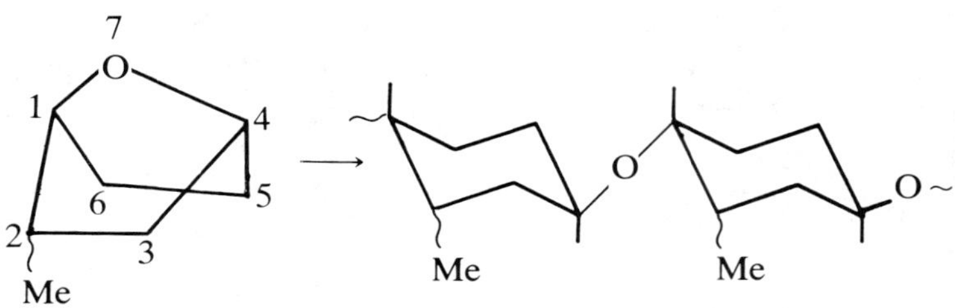

This means that the C-4 carbon atom selectively undergoes inversion of configuration.

The thermodynamic parameters that have been determined for the polymerization of bicyclic ethers are summarized in Table 3.2. Both the enthalpies and the entropies of polymerization are higher for the bicyclic ethers than for unsubstituted THF. However, $-\Delta H_p$ resulting from the strain released on ring opening more than compensates for the unfavorable increase in $-\Delta S^0_p$.

TABLE 3.1
Polymerizability of Bicyclic Ethers Containing a THF Ring

Name	Structure		Polymerizability[a]		Ref.
8-Oxabicyclo[4.3.0]nonane	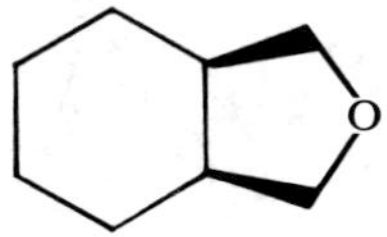	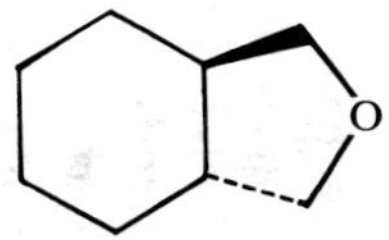	*cis* *trans*	–[b] +	127 127
7-Oxabicyclo[4.3.0]nonane		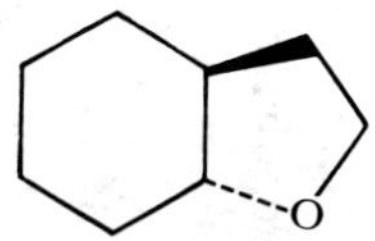	— *trans*	 +	128 128
3-Oxabicyclo[3.3.0]octane	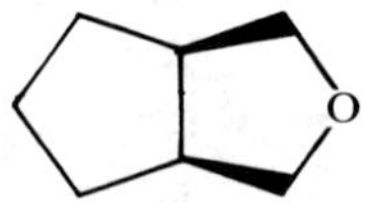	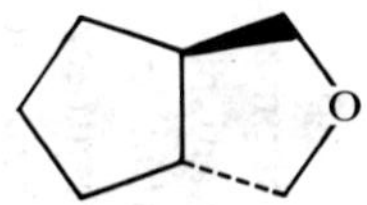	*cis* *trans*	– +	96 96, 129
2-Oxabicyclo[3.3.0]octane	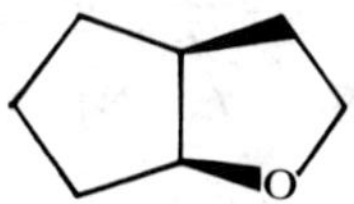	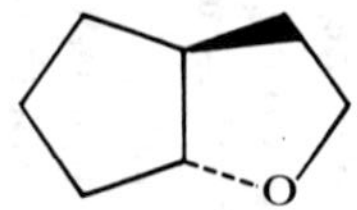	*cis* *trans*	– +	130 129, 130
3-Oxabicyclo[3.2.0]heptane	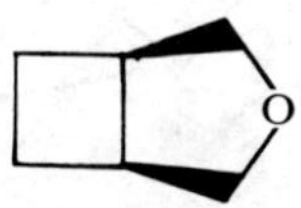		*cis* —	–	131

Compound	Structure			Polymerizability	References
3-Oxabicyclo[3.1.0]hexane	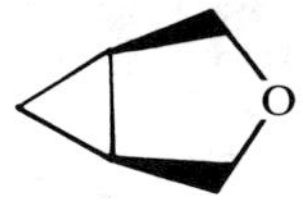			–	132
7-Oxabicyclo[2.2.1]heptane	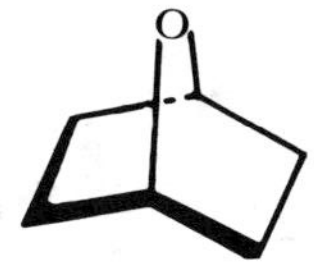			+	133, 134–8
2-Methyl-7-oxabicyclo[2.2.1]heptane	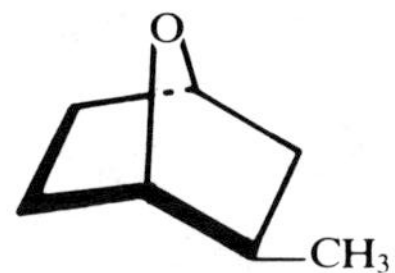	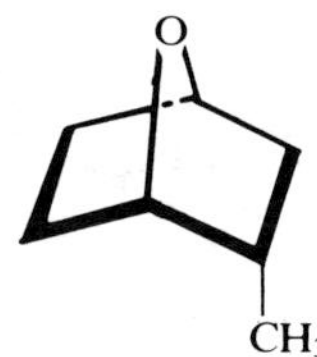	*exo*[c]	+	133, 136, 139, 140–6, 147
			endo[c]	+	133, 136, 139, 140, 143–8
2-Chloro-7-oxabicyclo[2.2.1]heptane	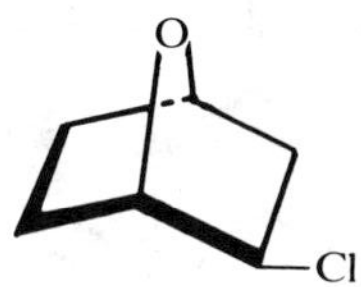			+	136, 149
2-*t*-Butyl-7-oxabicyclo[2.2.1]heptane	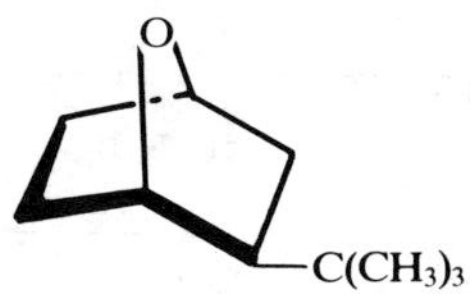	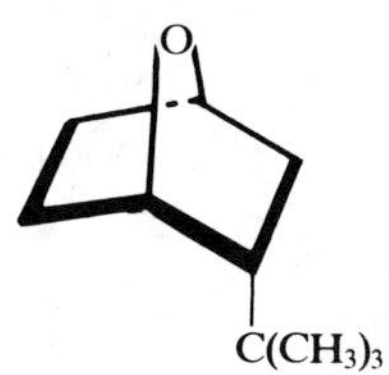		+	150, 151

TABLE 3.1 (*continued*)

Name	Structure	Polymerizability[a]		Ref.
2,6-Dimethyl-7-oxa-bicyclo[2.2.1]heptane	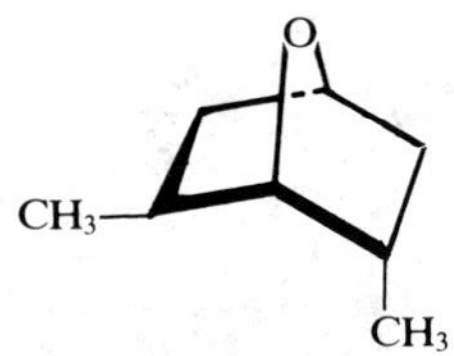	*endo, exo*	+	144
2,3-Dimethyl-7-oxa-bicyclo[2.2.1]heptane	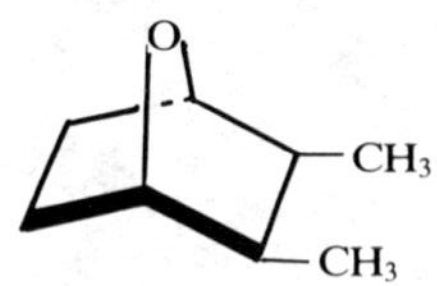	*exo, exo*	+	152
	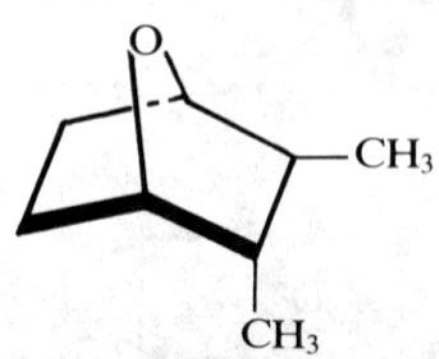	*endo, exo*	+	152

Monomer	Structure	Configuration		Ref.
	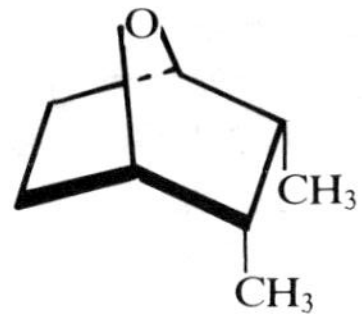	*endo, endo*	+	152
2,5-Dimethyl-7-oxa-bicyclo[2.2.1]heptane	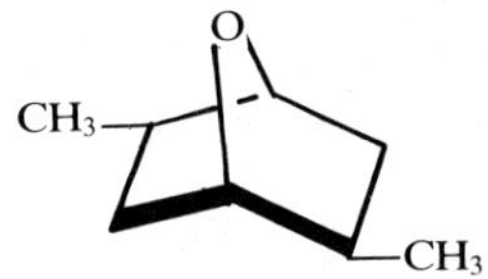	*exo, exo*	+	152
	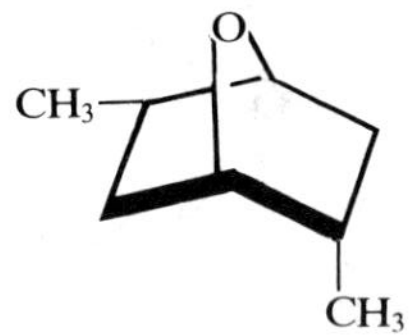	*endo, exo*	+	152
	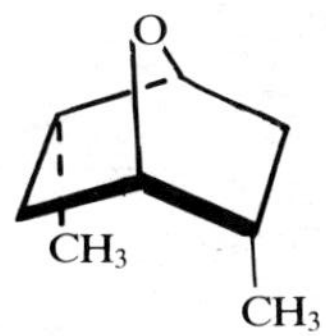	*endo, endo*	+	152

TABLE 3.1 (*continued*)

Name	Structure	Polymerizability[a]	Ref.
Fused cyclohexane-7-oxa-bicyclo[2.2.1]heptane	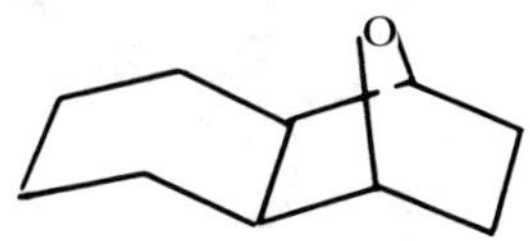	–	133
Fused bicycloheptane-7-oxa-bicyclo[2.2.1]heptane	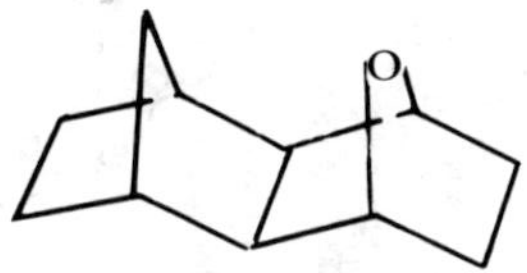	–	133
2,3-Benzo-7-oxa-bicyclo[2.2.1]heptane	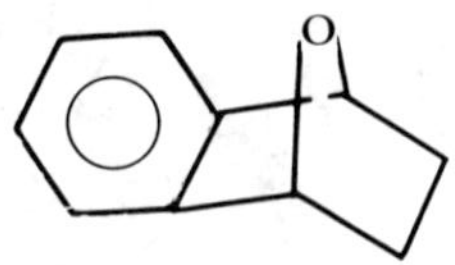	+	133

[a]A plus sign indicates that the monomer has been polymerized, and a minus sign that the monomer has not been homopolymerized.
[b]The *cis* monomer was copolymerized with the *trans* monomer.
[c]*Exo* structures are in the left column, *endo* structures are in the right column.

TABLE 3.2
Thermodynamic Parameters for Polymerization of Bicyclic Ethers in Comparison with THF[a]

Monomer	$-\Delta H_p$, kJ mol^{-1}	$-\Delta S_p^0$ J mol^{-1} K^{-1}	T_c, °C	Ref.
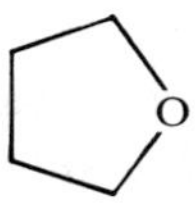	12.6	41.0	85	26
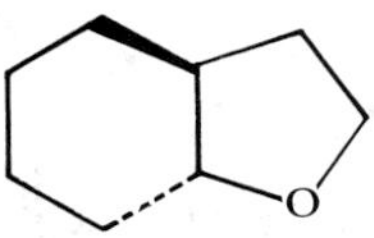	14.0	55.7	12	128
	44.3	75.3	320	139
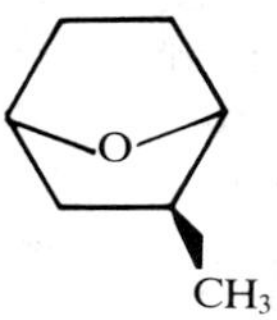	49.7	96.2	240	139
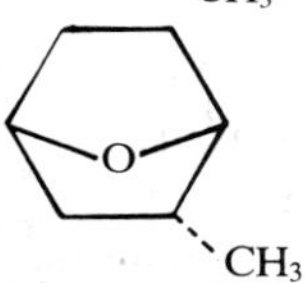	45.4	96.2	200	139

[a]The data are for liquid monomer → condensed polymer (bulk polymerizations).

3.4 COPOLYMERIZATION

3.4.1 General comments

In this section only copolymerization of THF with other monomers will be considered (reactions involving preformed

PTHF or other polymers will be discussed in Chapters 7 and 8), and only the salient factors affecting copolymerization will be mentioned. From these the reader will be able to discern that, although copolymerization of some monomers with THF does occur, one cannot always predict *a priori* the nature of the products that will form. Many more studies need to be made before even the nature of all the chemical reactions that occur are completely understood. For further details, the reader should consult the original papers mentioned in the discussion below, as well as earlier reviews.[1,31,153–4]

3.4.2 Copolymerizability

A few of the facts about the copolymerizability of THF with other monomers, already mentioned in the discussion of the initiation of THF polymerization (Section 3.2.1), are recapitulated below.

a) Clearly, since *cyclic trialkyloxonium* and *carboxonium ions* can add to THF to initiate its polymerization, copolymerization with monomers that can form such ions is expected, and occurs whenever the tertiary oxonium ion that forms from THF can react with those monomers to regenerate the initiating ion. Thus, THF readily copolymerizes with higher epoxides such as oxetanes,[26,155–9] 7-oxabicyclo[2.2.1]heptane,[133] dioxolane,[26,160–2] trioxane[163] and 4-methyl-1,3-dioxene-4.[164]

Many cyclic ethers that have little or no homopolymerizability will copolymerize with THF. Such monomers include 2-methyltetrahydrofuran,[26] 3-methyltetrahydrofuran[165] and 1,4-dioxane.[26] Random copolymerization does not always occur.[156,166] Sometimes the copolymers are contaminated by homopolymers which are difficult to distinguish and separate from the copolymers.[162,167] The effect of the counterion on the copolymerization of THF and 1,3-dioxolane is particularly worth noting.[162] Fractionation of the reaction products from copolymerizations in chloroform at 0°C and microstructural analysis by ^1H and ^{13}C NMR showed a marked dependence on the counterion. Products obtained with CH_3OSO_2F as initiator were real copolymers containing an appreciable amount of crossover diads, while those with Et_3OBF_4 as initiator were block copolymers and/or mixtures of homopolymers.[162] Consumption of monomers during copolymerization was followed by gas–liquid chromatography and showed trends consistent with the above interpretation. Both THF and 1,3-dioxolane are consumed concurrently from the initial stage of polymerizations with CH_3OSO_2F; but THF is consumed first when Et_3OBF_4 is used.[162]

Quantitative conversions of both monomers to copolymer can be obtained with many of these monomers.[26] Since some of the monomers, notably the oxetanes, also lead to "living" polymerizations, high-molecular-weight copolymers can be prepared under controlled conditions.

b) As can be inferred from the fact that sulfonium ions do not react with THF to form tertiary oxonium ions,[14] copolymerization with *cyclic sulfides* does not occur.

c) When a promoter reacts with THF and a Lewis acid to form a tertiary oxonium ion, the elements of the promoter add to THF. Thus, from the discussion in Section 3.2.1, one can predict that ***unsubstituted, monosubstituted*** and ***disubstituted 1,2-epoxides*** will copolymerize with THF, and they do.[39,156–7,168–77] The tertiary oxonium ion from THF is again able to initiate their copolymerization. *Trisubstituted 1,2-epoxides* apparently react with THF oxonium ions, but copolymerization does not occur, because the resulting oxonium ion does not initiate THF polymerization.[39] Hexafluoropropylene oxide ($CF_3\overline{CFOCF_2}$) also results in end-capped PTHF instead of random copolymer.[178] Since cationic polymerizations of 1,2-epoxides are prone to many side reactions—including transfer, termination and cyclic oligomer formation—copolymers with the oxiranes have lower molecular weights and are often oils, even though high conversion can be attained.

d) At ordinary polymerization temperatures, *carbenium ions* generally do not add to THF and, conversely, tertiary oxonium ions from THF do not initiate the cationic polymerization of common vinyl monomers such as styrene.[69] When polymer sequences from both monomers are present in the products, the latter are usually mixtures of homopolymers.[1] *Cyclopentadiene*, however, does appear to form true copolymers.[70–71] This is one way of introducing unsaturation into PTHF so that sulfur curing reactions can be carried out. A more common way is to copolymerize THF with *allyl glycidyl ether* (1-allyloxy-2,3-epoxy propane, $CH_2OCHCH_2OCH_2CH{=}CH_2$.[70,71,179,180] Other materials which have been used for the same purpose are *3-methyl-3-allyloxymethyloxetane*, $CH_2OCH_2C(CH_3)(CH_2OCH_2CH{=}CH_2)$;[159] *5,6-epoxy-1-hexene*, $CH_2OCH{-}CH_2CH{=}CH_2$;[179–80] *4,5-epoxy-1-hexene*, $CH_3CHOCH{-}CH_2CH{=}CH_2$;[177–80] *4-vinyl cyclohexene oxide*, $CH_2CH_2CHOCH{-}CH_2CH(CH{=}CH_2)$;[180] and *glycidyl methacrylate*, $CH_2OCHCH_2OC({=}O){-}C(CH_3){=}CH_2$.[181] Kuntz states that no evidence for gel formation was observed in copolymerizations with allyl glycidyl ether and $Ph_3C^+SbCl_6^-$ as initiator,[70,71] but Dreyfuss and Macey preferred the hexene oxides because, in their hands, copolymerizations with allyl glycidyl ether and triethyloxonium hexafluorophosphate as initiator gave only gelled product.[180] The reason for the difference may lie in the transfer and termination reactions that are known to occur with $SbCl_6^-$ counterions.

e) Copolymerizations with *lactones*[182] and with *cyclic anhydrides*[183] have also been reported. In my hands, side reactions occur, and further study is needed to verify that true copolymers are formed.

3.4.3 Copolymerization above T_c for THF homopolymerization

Copolymerization of THF will occur above the ceiling temperature for THF homopolymerization (see Section 3.1.3 for a discussion of T_c). In these cases polymers approaching alternating copolymers can form. In copolymerizations with 3,3-dimethyloxetane or 3,3-bis(fluoromethyl)oxetane, the limiting composition of the copolymer tended toward one mole of THF for each mole of oxetane when a large excess of THF was used. Kubisa and Penczek found that, if the comonomer was 3,3-bis(chloromethyl)oxetane, the limiting composition of the copolymer was close to two moles of THF for each mole of oxetane.[184] They ascribe the difference to the bulkiness of the $-CH_2Cl$ groups compared to $-CH_3$ or $-CH_2F$ groups, which results in lack of freedom of rotation of the units containing the $-CH_2Cl$ groups. Thus the influence of the penultimate unit during propagation or of the antepenultimate unit in the depropagation step cannot be ignored.

3.4.4 Factors that influence copolymerization

All the problems inherent in homopolymerizations of THF are encountered in copolymerizations also. In addition, new problems result from the simultaneous presence of two polymerizable monomers and their interaction with each other. The nature and number of active centers generated in a given copolymerization are even more uncertain than in homopolymerizations, because there are additional transfer- and termination reaction possibilities. Solvent effects still exist,[175,185] but in some cases copolymerizations can be carried to 100% conversion based on both monomers.[26] The thermodynamic considerations controlling conversions in THF homopolymerizations do not apply to copolymerizations. Backbiting and effects of penultimate units are possible.[184,186] Current evidence suggests that the two factors which have the greatest effect on the course of copolymerizations of THF with other monomers are relative basicity and ring strain in the monomers. The basicity of cyclic compounds containing oxygen is affected by chemical structure, ring size and substituents. The effects of cyclic-ether ring size on basicity are in the order: $4>5>6>3$. In five-membered rings the basicity order is: ether > lactone > formal. Methyl substituents increase basicity, while chloromethyl substituents decrease it. The reactivities of the monomers in copolymerization experiments are broadly in the same order as their basicities,[153] but ringstrain also has an influence. Tanaka has derived equations relating the relative contributions of ring strain and basicity to the reactivity of cyclic ethers in cationic copolymerizations. He finds good agreement between predicted and observed reactivities.[187] In addition, kinetic factors also seem to play an important role in copolymerizations of THF. Some of these factors will be discussed in Chapter 4.

REFERENCES

1. P. Dreyfuss and M. P. Dreyfuss, *Adv. Polym. Sci.*, **4**, 528 (1967).
2. H. Meerwein, D. Delfs and H. Morshel, *Angew. Chem.*, **72**, 927 (1960).
3. J. Furukawa and T. Saegusa, "Polymerization of Aldehydes and Oxides", Interscience, New York, NY (1963).
4. T. Saegusa and S. Kobayashi, *Prog. Polym. Sci. Jap.*, **6**, 107 (1973).
5. N. S. Enikolopian, V. V. Ivanov, G. V. Koronina, M. A. Markevich, T. I. Prokof'eva, A. T. Ponomarenko and G. V. Rakova, *Vysokomol. Soyed.* **A19**, 1924 (1977) (in Russian). Translated in *Polym. Sci. USSR*, **19**, 2200 (1977).
6. B. A. Rosenberg, W. I. Irshak and N. S. Enikolopian, "Interchain Reactions in Polymeric Systems", Khimiya, Moscow, USSR (1975).
7. H. Meerwein, G. Hinz, P. Hoffman, E. Kroning and E. Pfeil, *J. Prakt. Chem.*, **147**, 257 (1937).
8. Y. Eckstein and P. Dreyfuss, *J. Inorg. Nucl. Chem.*, **43**, 23 (1981).
9. S. Nakahama, S. Hino and N. Yamazaki, *Polym. J.*, **2**, 56 (1971).
10. C. F. Heins, *J. Polym. Sci. B*, **7**, 625 (1969).
11. N. Yamazaki, *Adv. Polym. Sci.*, **6**, 377 (1969).
12. S. N. Bhadani, PhD Thesis, University of Manitoba, p. 120 (1966).
13. Y. Eckstein and P. Dreyfuss, *J. Polym. Sci.: Polym. Chem. Ed.*, **17**, 4115 (1979).
14. J. L. Lambert and E. J. Goethals, *Makromol. Chem.*, **133**, 289 (1970).
15. J. V. Crivello and J. H. W. Lam, *J. Polym. Sci.: Polym. Chem. Ed.*, **17**, 977 (1979).
16. Y. Takegami, T. Ueno and R. Hirai, *J. Polym. Sci.: A1*, **4**, 973 (1966).
17. E. L. Muetterties, US Patent 2,856,370 (Oct. 1958).
18. E. L. Muetterties, T. A. Bither, M. W. Farlow and D. D. Coffman, *J. Inorg. Nucl. Chem.*, **16**, 52 (1960).
19. J. V. Crivello and J. H. W. Lam, *Macromolecules*, **10**, 1307 (1977).
20. J. V. Crivello and J. H. W. Lam, *J. Polym. Sci.: Polym. Lett. Ed.*, **16**, 563 (1978).
21. G. H. Smith, Belgian Patent 828,841 (Nov. 7, 1975).
22. Belgian Patent 837,782, to Imperial Chemicals Industries (June 22, 1976).
23. J. H. Binks and M. B. Huglin, *Makromol. Chem.*, **93**, 268 (1966).
24. Y. Takegami, T. Ueno and R. Hirai, *Bull. Chem. Soc. Japan*, **38**, 1222 (1965).

25. J. V. Crivello and J. H. W. Lam, *J. Polym. Sci.: Polym. Chem. Ed.*, **17**, 1047 (1979).
26. M. P. Dreyfuss and P. Dreyfuss, *J. Polym. Sci.: A1*, **4**, 2179 (1966).
27. F. S. Dainton and K. Ivin, *Quart. Rev.*, **12**, 61 (1958).
28. C. Huang and H. Wang, *J. Polym. Sci.: A*1, **10**, 791 (1972).
29. H. Sawada, *J. Macromol. Sci., Revs. Macromol. Chem.*, **C8**, 235 (1972).
30. K. J. Ivin and J. Leonard, *Polymer*, **6**, 621 (1965).
31. P. Dreyfuss and M. P. Dreyfuss, in "Comprehensive Chemical Kinetics", Vol. 15, C. H. Bamford and C. F. H. Tipper (eds.), Elsevier, Amsterdam-W, Netherlands, Chap. 4, pp. 290–1 (1976).
32. A. M. Buyle, K. Matyjaszewski and S. Penczek, *Macromolecules*, **10**, 269 (1977).
33. H. Meerwein, German Patent 741,478 (June 21, 1939).
34. H. Meerwein, D. Delfs and H. Morshel, *Angew. Chem.*, **72**, 927 (1960).
35. Y. Yamashita, S. Kozawa, M. Hirota, K. Chiba, H. Matsui, A. Hirao, M. Kadama and K. Ito, *Makromol. Chem.*, **142**, 171 (1971).
36. T. Saegusa and S. Kobayashi, *ACS Symposium Series*, **6**, 150 (1975).
37. E. Franta, L. Reibel, J. Lehmann and S. Penczek, *J. Polym. Sci.: Symp.*, **56**, 139 (1976).
38. F. Afshar-Taromi, M. Scheer, P. Rempp and E. Franta, *Makromol. Chem.*, **179**, 849 (1978).
39. P. Dreyfuss and J. P. Kennedy, *J. Appl. Polym. Sci.: Appl. Polym. Symp.*, **30**, 165 (1977).
40. R. W. Campbell and O. Vogl, *J. Macromol. Sci.-Chem.*, **A11**, 515 (1977).
41. T. Saegusa, H. Imai, S. Hirai and J. Furukawa, *Makromol. Chem.*, **54**, 218 (1962).
42. P. Dreyfuss and J. P. Kennedy, *J. Polym. Sci.: Polym. Symp.*, **60**, 47 (1977).
43. R. Hoene and K.-H. W. Reichert, *Makromol. Chem.*, **177**, 3545 (1976).
44. R. A. Goodrich and P. M. Treichel, *J. Am. Chem. Soc.*, **88**, 3509 (1966).
45. D. Sims, *Makromol. Chem.*, **98**, 245 (1966).
46. R. C. Burrows and B. F. Crowe, *J. Appl. Polym. Sci.*, **6**, 465 (1962).
47. H. Meerwein and K. Wunderlich, *Angew. Chem.*, **69**, 481 (1958).
48. J. B. Lambert and D. H. Johnson, *J. Am. Chem. Soc.*, **90**, 1349 (1968).
49. P. Dreyfuss and J. P. Kennedy, *J. Polym. Sci.: Symp.*, **56**, 129 (1976).
50. E. Franta, F. Afshar-Taromi and P. Rempp, *Makromol. Chem.*, **177**, 2191 (1976).
51. F. J. Burgess, A. V. Cunliffe, D. H. Richards and D. C. Sherrington, *J. Polym. Sci.: Polym. Lett. Ed.*, **14**, 471 (1976).
52. H. Meerwein and K. Wunderlich, *Angew. Chem.*, **69**, 481 (1957).
53. J. B. Lambert and D. H. Johnson, *J. Am. Chem. Soc.*, **90**, 1349 (1968).
54. K. I. Lee and P. Dreyfuss, *ACS Symposium Series*, **59**, 24 (1977).
55. J. Lehmann and P. Dreyfuss, *Adv. Chem. Series*, **176**, 587 (1979).
56. G. A. Olah, B. G. B. Gupta and S. C. Narang, *Synthesis*, 274 (1979).
57. H. J. Breunig and D. Wöhrle, *Makromol. Chem.*, **179**, 1653 (1978).

58. P. Dreyfuss and M. P. Dreyfuss, *Adv. Chem. Series*, **91**, 335 (1969).
59. T. Saegusa and S. Matsumoto, *J. Macromol. Sci.-Chem.*, **A4**, 873 (1970).
60. Y. Yamashita, S. Kozawa, M. Hirota, K. Chiba, H. Matsui, H. Hirao, M. Kadama and K. Ito, *Makromol. Chem.*, **142**, 171 (1971).
61. Y. Yamashita, H. Matsui, G. Hattori, S. Kozawa and M. Hirota, *Makromol. Chem.*, **142**, 183 (1971).
62. G. Pruckmayr and T. K. Wu, *Macromolecules*, **11**, 265, 662 (1978).
63. M. P. Dreyfuss, J. C. Westfahl and P. Dreyfuss, *Macromolecules*, **1**, 437 (1968).
64. C. E. H. Bawn, R. M. Bell and A. Ledwith, *Polymer*, **6**, 95, 611 (1965).
65. I. Kuntz, *J. Polym. Sci.*: *A*1, **5**, 193 (1967).
66. A. Ledwith, *Adv. Chem. Series*, **91**, 317 (1969).
67. W. A. Pasika, *J. Polym. Sci.*: *A1*, **7**, 1489 (1969).
68. F. Klages, J.E. Gordon and H. A. Jung, *Chem. Ber.*, **98**, 3748 (1965).
69. S. Aoki, Y. Harita, T. Otsu and M. Imoto, *Bull. Chem. Soc. Japan* **38**, 1922 (1965).
70. I. Kuntz and R. N. Thomas, US Patent 3,435,015 (to Esso Research and Engineering Co.) (March 25, 1969); CA **70**, 10756 p (1969).
71. I. Kuntz, *Trans. NY Acad. Sci.*, **33**, 529 (1971).
72. G. A. Olah and T.-L. Ho, *Synthesis*, 609 (1976).
73. T.-L. Ho and G. A. Olah, *J. Org. Chem.*, **42**, 3097 (1977).
74. F. A. M. Abdul-Rasoul, A. Ledwith and Y. Yağci, *Polym. Bull.*, **1**, 1 (1978).
75. H. R. Ward and P. D. Sherman, Jr., *J. Am. Chem. Soc.*, **90**, 3812 (1968).
76. J. V. Crivello and J. H. W. Lam, *J. Polym. Sci.*: *Polym. Chem. Ed.*, **17**, 1059 (1979).
77. M. E. Woodhouse, F. D. Lewis and T. J. Marks, *J. Am. Chem. Soc.*, **100**, 996 (1978).
78. H. Yoshida and K. Hayashi, *Adv. Polym. Sci.*, **6**, 401 (1969).
79. N. Oguni, M. Kamachi and J. K. Stille, *Macromolecules*, **7**, 435 (1974).
80. J. M. Sangster and D. J. Worsfold, *Macromolecules*, **5**, 229 (1972).
81. J. M. Sangster and D. J. Worsfold, *J. Macromol. Sci.-Chem.*, **A7**, 1415 (1973).
82. M. Szwarc, "Carbanions, Living Polymers, and Electron Transfer Processes", Interscience, New York, NY, Chap. VII, p. 405 (1968).
83. S. Smith and A. J. Hubin, UK Patent 1,120,304 (to Minnesota Mining and Manufacturing Co.) (July 17, 1968).
84. A. J. Hubin and S. Smith, US Patent 3,436,359 (to Minnesota Mining and Manufacturing Co.) (April 1, 1969).
85. S. Smith and A. J. Hubin, *J. Macromol. Sci.-Chem.*, **A7**, 1399 (1973).
86. T. Saegusa and S. Kobayashi, *ACS Symposium Series*, **6**, 150 (1975).
87. G. Pruckmayr and T. K. Wu, *Macromolecules*, **6**, 33 (1973).
88. T. K. Wu and G. Pruckmayr, *Macromolecules*, **8**, 77 (1975).
89. K. Matyjaszewski, P. Kubisa and S. Penczek, *J. Polym. Sci.*: *Polym. Chem. Ed.*, **12**, 1333 (1974).
90. K. Matyjaszewski and S. Penczek, *Macromolecules*, **7**, 173 (1974).

91. R. J. Gillespie, *Accounts Chem. Res.*, **1**, 202 (1968).
92. R. J. Gillespie and E. A. Robinson, "Non-aqueous Solvent Systems", T. C. Waddington (ed.), Academic, London, p. 117 (1965).
93. A. Hase and T. Hase, *Analyst*, **97**, 998 (1972).
94. T. Saegusa and S. Matsumoto, *J. Polym. Sci.*: *A1*, **6**, 1559 (1968).
95. T. Saegusa, Kyoto University, personal communication (1976).
96. J. Kops, S. Hvilsted and G. Sorensen, in "Proceedings of First International Symposium on Polymerization of Heterocycles, Ring-opening", Warsaw-Jablonna, p. 132 (1975).
97. S. Hvilsted, Licentiatus Technices Thesis, Instituttet for Kemiindustri, Denmark, p. 20 (1978).
98. D. P. Lee and P. Dreyfuss, *J. Polym. Sci.*: *Polym. Chem. Ed.*, **18**, 1627 (1980).
99. S. Penczek, Polish Academy of Sciences, Łódź, Poland, personal communication (1978).
100. K. Brzezińska, W. Chwiałkowska, P. Kubisa, K. Matyjaszewski and S. Penczek, *Makromol. Chem.*, **178**, 2491 (1977).
101. Y. Eckstein, D. P. Lee, R. P. Quirk and P. Dreyfuss, *J. Polym. Sci.*: *Polym. Chem. Ed.*, **18**, 2021 (1980).
102. Y. Eckstein and P. Dreyfuss, *Anal. Chem.*, **52**, 537 (1980).
103. S. Smith and A. J. Hubin, US Patent 3,824,197 (to Minnesota Mining and Manufacturing Co.) (July 16, 1974).
104. G. Berger, M. Levy and D. Vofsi, *J. Polym. Sci.*, **B4**, 183 (1966).
105. Y. Yamashita, K. Nobutoki, Y. Nakamura and M. Hirota, *Macromolecules*, **4**, 548 (1971).
106. F. J. Burgess, A. V. Cunliffe, J. V. Dawkins and D. H. Richards, *Polymer*, **18**, 733 (1977).
107. R. A. Barzykina, G. N. Komratov, G. V. Korovina and S. G. Entelis, *Vysokomol. Soyed*, **A20**, 170 (1978) (in Russian). Translated in *Polym. Sci. USSR*, **20**, 197 (1978).
108. G. N. Komratov, R. A. Barzykina and G. V. Korovina, *Vysokomol. Soyed*, **A20**, 608 (1978) (in Russian). Translated in *Polym. Sci. USSR*, **20**, 686 (1979).
109. T. Saegusa and S. Matsumoto, *J. Macromol. Sci.*, **A4**, 873 (1970).
110. S. Penczek, J. Fejgin, P. Kubisa, K. Matyjaszewski and M. Tomaszewicz, *Makromol. Chem.*, **172**, 243 (1973).
111. G. Pruckmayr and T. K. Wu, *Macromolecules*, **11**, 662 (1978).
112. A. Miyake and W. H. Stockmayer, *Makromol. Chem.*, **88**, 90 (1965).
113. B. A. Rozenberg, E. B. Lyudvig, A. R. Gantmakher and S. S. Medvedev, *Vysokomol. Soyed.*, **6**(11), 2035 (1964) (in Russian). Translated in *Polym. Sci. USSR*, **6**(11) 2253 (1964).
114. E. J. Goethals, *Adv. Polym. Sci.*, **23**, 103 (1977).
115. J. Dale, G. Borgen and K. Daasvatn, *Acta. Chem. Scand.*, *Ser. B*, **28**, 378 (1974).
116. J. B. Rose, *J. Chem. Soc.*, 542 (1956).

117. P. Dreyfuss and M. P. Dreyfuss, *Polym. J.*, **8**, 81 (1976).
118. D. Van Ooteghem and E. J. Goethals, *Makromol. Chem.*, **175**, 1513 (1974).
119. J. L. Lambert, D. Van Ooteghem and E. J. Goethals, *J. Polym. Sci.: A1*, **9**, 3055 (1971).
120. J. M. McKenna, T. K. Wu and G. Pruckmayr, *Macromolecules*, **10**, 877 (1977).
121. G. Pruckmayr and T. K. Wu, *Macromolecules*, **11**, 265 (1978).
122. F. S. Dainton, T. R. E. Devlin and P. A. Small, *Trans. Faraday Soc.*, **51**, 1710 (1955).
123. P. A. Small, *Trans. Faraday Soc.*, **51**, 1717 (1955).
124. H. K. Hall, Jr., *Polymer Preprints*, **6**(2), 535 (1965).
125. R. Chiang and J. H. Rhodes, *J. Polym. Sci. B*, **7**, 643 (1969).
126. J. B. Rose, in "The Chemistry of Cationic Polymerization", P. H. Plesch (ed.), Macmillan, New York, NY, p. 433 (1963).
127. J. Kops and H. Spanggaard, *Makromol. Chem.*, **175**, 3077 (1974).
128. J. Kops, E. Larsen and H. Spanggaard, *J. Polym. Sci. C*, **56**, 91 (1976).
129. S. Hvilsted and J. Kops, *Macromolecules*, **12**, 889 (1979).
130. S. Hvilsted, Licentiatus Technices Thesis, Instituttet for Kemmindustri, Denmark, p. 11 (1978).
131. K. C. Steuben, *J. Polym. Sci.: A1*, **4**, 829 (1966).
132. T. Shono, A. Oku, T. Morikawa, M. Kimura and R. Oda, *Bull. Chem. Soc. Japan*, **38**, 940 (1965).
133. E. L. Wittbecker, H. K. Hall, Jr. and T. W. Campbell, *J. Am. Chem. Soc.*, **82**, 1218 (1960).
134. K. Weissermel and E. Nölken, *Makromol. Chem.*, **68**, 140 (1963).
135. P. Giusti and F. Andruzzi, *Ann. Chim. (Rome)*, **56**, 973 (1966).
136. M. Baccaredda, P. Giusti, F. Andruzzi, P. Cerrai and G. Masetti, *Chim. Ind. (Milan)*, **50**, 81 (1968).
137. P. Giusti, F. Andruzzi and M. Di Maina, *Chim. Ind. (Milan)*, **50**, 1213 (1968).
138. J. Stejny, *J. Macromol. Sci.-Chem.*, **A7**, 1435 (1973).
139. F. Andruzzi, G. Pilcher, Y. Virmani and P. H. Plesch, *Makromol. Chem.*, **178**, 2367 (1977).
140. P. L. Magagnini, F. Sardelli, F. Andruzzi and P. Giusti, *Ann. Chim. (Rome)*, **57**, 1493 (1967).
141. P. Giusti, U. Fiorentino, G. Turchi, F. Andruzzi and P. L. Magagnini, *Makromol. Chem.*, **128**, 1 (1969).
142. M. Baccaredda, P. Giusti, F. Andruzzi, P. Cerrai and M. Di Maina, *J. Polym. Sci. C*, **31**, 157 (1970).
143. J. Kops and H. Spanggaard, *Polymer Preprints*, **13**, 90 (1972).
144. J. Kops and H. Spanggaard, *Makromol. Chem.*, **151**, 21 (1972).
145. T. Saegusa, M. Motoi, S. Matsumoto and H. Fujii, *Macromolecules*, **5**, 233 (1972).
146. T. Saegusa, S. Matsumoto, M. Motoi and H. Fujii, *Macromolecules*, **5**, 236 (1972).

147. J. Kops and H. Spanggaard, *J. Macromol. Sci.-Chem.*, **A7**, 1455 (1973).
148. T. Saegusa, M. Motoi, S. Matsumoto and H. Fujii, *Macromolecules*, **5**, 815 (1972).
149. F. Andruzzi, P. Cerrai and P. Giusti, *Ann. Chim.* (*Rome*), **59**, 91 (1969).
150. T. Saegusa, M. Motoi and H. Suda, *Macromolecules*, **9**, 231 (1976).
151. T. Saegusa, M. Motoi and H. Suda, *Macromolecules*, **9**, 526 (1976).
152. T. Saegusa, M. Motoi, H. Suda, H. Kasai and A. Ueyama, *Polym. J.*, **11**, 113 (1979).
153. P. Dreyfuss and M. P. Dreyfuss, in "Kinetics and Mechanisms of Polymerization Reactions", Vol 2. "Polymerization by Ring-opening Reactions", K. C. Frisch and S. L. Reegen (eds.), Marcel Dekker, New York, NY pp. 111–58 (1969).
154. P. Dreyfuss and M. P. Dreyfuss, in *Encyclopedia of Polymer Science and Technology*, Vol. 13, H. Mark and N. G. Gaylord (eds.), Wiley, New York, NY, pp. 670–91 (1970).
155. T. Saegusa, H. Imai and J. Furukawa, *Makromol. Chem.*, **56**, 55 (1962).
156. T. Saegusa, T. Ueshima, H. Imai and J. Furukawa, *Makromol. Chem.*, **79**, 221 (1964).
157. B. A. Rozenberg, E. B. Lyudvig, N. V. Destyatova, A. R. Gantmakher and S. S. Medvedev, *Vysokomol. Soed.*, **7**(6), 1010 (1965) (in Russian). Translated in *Polym. Sci. USSR*, **7**(6), 1116 (1965).
158. I. Penczek and S. Penczek, *Polymer Lett.*, **5**, 367 (1967).
159. Yu. A. Gorin, E. I. Rodina, N. V. Kozlova and K. V. Nel'son, *Vysokomol. Soyed.*, **A11**(7), 1477 (1969) (in Russian). Translated in *Polym. Sci. USSR*, **11**(7), 1675 (1969).
160. M. Okada, N. Takikawa, S. Iwatsuki, Y. Yamashita and Y. Ishii, *Makromol. Chem.*, **82**, 16 (1965).
161. P. Kubisa and S. Penczek, *Macromolecules*, **10**, 1216 (1977).
162. Y. Yokoyama, M. Okada and H. Sumitomo, *Polym. J.*, **11**, 629 (1969).
163. M. B. Price and F. B. McAndrew, *Polymer Preprints*, **7**(1), 207 (1966).
164. W. Kawai, *J. Polym. Sci.: A1*, **6**, 137 (1968).
165. H. Stratmann, B. Stützel and R. Feinauer, *Angew. Makromol. Chem.*, **74**, 105 (1978).
166. K. Tada, T. Saegusa and J. Furukawa, *Makromol. Chem.*, **71**, 71 (1964).
167. S. Aoki, Y. Harita, T. Otsu and M. Imoto, *Bull. Chem. Soc. Japan*, **38**, 1922 (1965).
168. W. J. Murbach and A. Adicoff, *Ind. Eng. Chem.*, **52**, 772 (1960).
169. L. A. Dickinson, *J. Polym. Sci.*, **58**, 857 (1962).
170. T. Saegusa, T. Kobayashi, S. Kobayashi, S. L. Couchman and O. Vogl, *Polym. J.*, **11**, 463 (1979).
171. J. M. Hammond, J. F. Hooper and W. G. P. Robertson, *J. Polym. Sci.: A1*, **9**, 265 (1971).
172. J. M. Hammond, J. F. Hooper and W. G. P. Robertson, *J. Polym. Sci., A1*, **9**, 281 (1971).
173. L. P. Blanchard, S. Kondo, J. Moinard, J. F. Pierson and F. Tahiani, *J. Polym. Sci.: A1*, **10**, 399 (1972).

174. E. J. Alvarez, V. Hornof and L. P. Blanchard, *J. Polym. Sci.*: *A1*, **10**, 1895 (1972).
175. E. J. Alvarez, V. Hornof and L. P. Blanchard, *J. Polym. Sci.*: *A1*, **10**, 2237 (1972).
176. L. P. Blanchard, J. C. Asselin, S. L. Malhotra, *Polym. J.*, **7**, 326 (1974).
177. L. P. Blanchard, G. G. Gabra and S. L. Malhotra, *J. Polym. Sci.*: *Polym. Chem. Ed.*, **13**, 1619 (1975).
178. T. G. Shchibriya, A. V. Ignatenko, S. P. Krukovskii and V. A. Ponomarenko, *Vysokomol. Soyed.*, **A18**(8), 1805 (1976) (in Russian). Translated in *Polym. Sci. USSR*, **A18**, 2065 (1976).
179. J. G. Burt and H. C. Walter US Patent 3,287,330 (to E. I. du Pont de Nemours and Co.) (Nov. 22, 1966).
180. M. P. Dreyfuss and J. H. Macey, *Polymer Preprints*, **20**(2), 324 (1979).
181. T. Otsu, K. Goto, S. Aoki and M. Imoto, *Makromol. Chem.*, **71**, 150 (1964).
182. Y. Yamashita, T. Tsuda, M. Okada and S. Iwatsuki, *J. Polym. Sci.*: *A1*, **4**, 2121 (1966).
183. W. Öchsner and K.-H. Reichert, *Makromol. Chem.*, **150**, 1 (1971).
184. P. Kubisa and S. Penczek, *J. Macromol. Sci.-Chem.*, **A7**, 1509 (1973).
185. G. V. Korovina, D. D. Novikov, D. Ya. Rossina, T. V. Grenevich, N. G. Taganov, R. A. Barzykina, G. N. Komratov and S. G. Entelis, *Vysokomol. Soyed*, **A18**(6), 1253 (1976) (in Russian). Translated in *Polym. Sci. USSR*, **18**, 1438 (1976).
186. Y. Yamashita, F. Inoue, G. Hattori and K. Ito, *Makromol. Chem.*, **151**, 91 (1972).
187. Y. Tanaka, *J. Macromol. Sci.-Chem.*, **1**, 1059 (1967).

Note added in proof: Since this book was written, another related review article has appeared. See:
S. Penczek, P. Kubisa and K. Matyjaszewski, *Adv. Polym. Sci.*, **37**, 1–151 (1980).

CHAPTER 4

Kinetics of Polymerization and Copolymerization

Very many studies of the kinetics of polymerization of THF have been carried out. There is general agreement among investigators in the field about the mechanism of polymerization. Kinetic parameters have been determined with a variety of counterions. The controversies that currently exist center around details such as the reasons for the observed effect of the dielectric constant of the polymerization media on specific rate constants.[1,2]

Except for some recent studies involving macroesters, kinetic studies of THF have been reviewed repeatedly and more frequently than any other aspect of THF polymerization.[3–13] In this chapter, results from more complete studies and from recent studies not included in previous reviews will be emphasized. The objective is to give the reader a feeling for the factors that are important in controlling the various rate constants, as well as an idea of the magnitude of the kinetic parameters under a variety of conditions. A more comprehensive and historical overview may be obtained by consulting the earlier reviews.

4.1 THE GENERAL KINETIC EQUATION

The polymerization of THF is represented formally by the scheme shown on page 76. Additional reactions that must be included when ion–ester equilibria can occur will be presented later in the chapter.

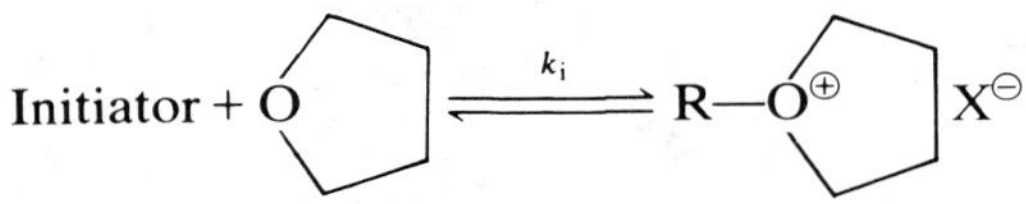

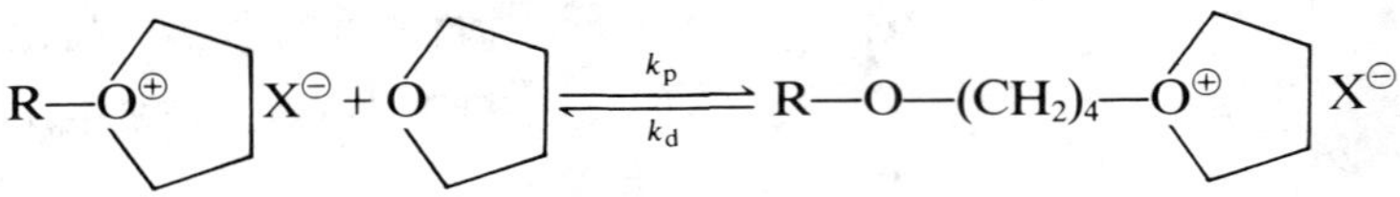

The many substances that have been added to THF to induce its polymerization have been described in Chapter 3. For the purposes of our initial discussion, the term initiator, I, is reserved solely for those substances which add directly to THF monomer to give *tertiary* THF oxonium ions, the propagating species. As noted already, many substances that induce the polymerization of THF do so indirectly and often inefficiently. In such cases kinetic analyses require separate determinations of the number and rate of production of oxonium ions. In more recent kinetic analyses of THF polymerization, separate determinations of the number of growing centers is standard practice, and the kinetic equations are modified accordingly. Methods for determining the number of growing centers formed during a THF polymerization were given in Section 3.2.3.

In the equations, k_i, k_p and k_d are the specific rate constants of initiation, propagation and depropagation, respectively. In most cases it is assumed that termination reactions are not important under the conditions studied. With many counterions at temperatures well below the ceiling temperature of THF polymerizations, these assumptions are entirely valid, especially at the low conversions where kinetic analyses are usually made. At higher temperatures, such as 50°C for $PF_6^{\ominus}$, $SbF_6^{\ominus}$ or $AsF_6^{\ominus}$ counterions or even at 0°C for $BF_4^{\ominus}$ and many other counterions, termination reactions cannot be ignored.

The kinetics of THF polymerization are expressed in terms of the rate of reaction for an equilibrium polymerization without termination:

$$-d[M]/dt = k_i[I]_t([M]_t - [M]_e) + k_p([I]_0 - [I]_t)([M]_t - [M]_e).$$

Here, $[M]_t$ is the monomer concentration at time t; $[I]_t$ is the initiator concentration at time t; $[M]_e$ is the equilibrium monomer concentration; and $[I]_0$ is the initiator concentration at $t = 0$. As long as initiation is fast and the initial rate is directly proportional to $[I]_0$, $[I]_t$ will be very small. The rate is then closely approximated by

$$-d[M]/dt = k_p[I]_0([M] - [M]_e)\,.$$

Of course, when the number of active centers is measured directly, $[I]_0$ becomes equal to the concentration of active centers, and the simplified form applies directly. After integration, the simplified equation has the form

$$k_p t = \frac{1}{[I]_0} \ln \frac{[M]_0 - [M]_e}{[M]_t - [M]_e}\,.$$

This is the usual form to which the experimental data are fitted. Sometimes, as in the work of Saegusa and Matsumoto,[14] initiation is not assumed to be instantaneous and the change in the number of active centers with time is taken into account. The equation then becomes on integration

$$\ln \frac{[M]_{t_1} - [M]_e}{[M]_{t_2} - [M]_e} = k_p \int_{t_1}^{t_2} [P^*]\, dt\,.$$

Here $[M]_{t_1}$ and $[M]_{t_2}$ are the monomer concentrations at times t_1 and t_2, respectively, and $[P^*]$ is the experimentally determined concentration of tertiary oxonium ions at a given time, t.

4.2 EFFECT OF INITIATOR

General Comments

The species which initiates THF polymerization is comprised of a cation and an anion. The nature of both portions greatly affects subsequent kinetic measurements.

The nature of the counterion X^- and the temperature at which the polymerization is carried out are important. For example, in a study of THF polymerizations at 30°C initiated by equivalent amounts of triethyloxonium salts with different counterions, Dreyfuss and Dreyfuss[15] observed differences in both conversion and rates of change in intrinsic viscosity with time of polymerization. In the cases of BF_4^- and $SbCl_6^-$, the final viscosities and conversions that were attained were lower than when SbF_6^- or PF_6^- counterions were used. The viscosity of BF_4^- polymers remained constant after constant conversion was reached, and so termination can be inferred. The viscosity of $SbCl_6^-$ polymers continued to decrease even after constant conversion was attained. With the $SbCl_6^-$ counterion, both termination and transfer occurred.

Earlier Vofsi and Tobolsky,[16] in a study with a $BF_4^{\ominus}$ counterion at 0°C, had inferred that termination must be considered at longer times.

In a comparison of rates of THF polymerization at 0°C initiated by $Et_3O^{\oplus}BF_4^{\ominus}$ and $Et_3O^{\oplus}AlCl_4^{\ominus}$, Saegusa and Matsumoto[17] confirmed the termination inferred by Dreyfuss and Dreyfuss[15] and by Vofsi and Tobolsky.[16] The termination reaction was less obvious at 0°C than at 30°C; nevertheless, it was clearly evident. Further, they found that termination with $AlCl_4^-$ counterion is even more serious. The comparison of the [P*] curves is shown in Figure 4.1.

Other studies of the stability of complex ions in cationic polymerization have been made, but these have focused primarily on triphenylmethylcarbenium ion salts, $(C_6H_5)_3C^{\oplus}X^{\ominus}$, and dioxolenium salts $\overline{OCH_2CH_2OCH}^{\oplus}X^{\ominus}$.[18–20] The conclusions that can be inferred relative to THF polymerizations are not substantially different from those already described.

One further comment about the stability of counterion–oxonium-ion complexes is in order. Whereas maximum stability of the counterion itself under polymerization conditions is desirable, intermediate stability of the complex may be preferable. The complex should not be so stable that polymerization or desired termination reactions are inhibited. Initiator fragments that are difficult to remove quantitatively

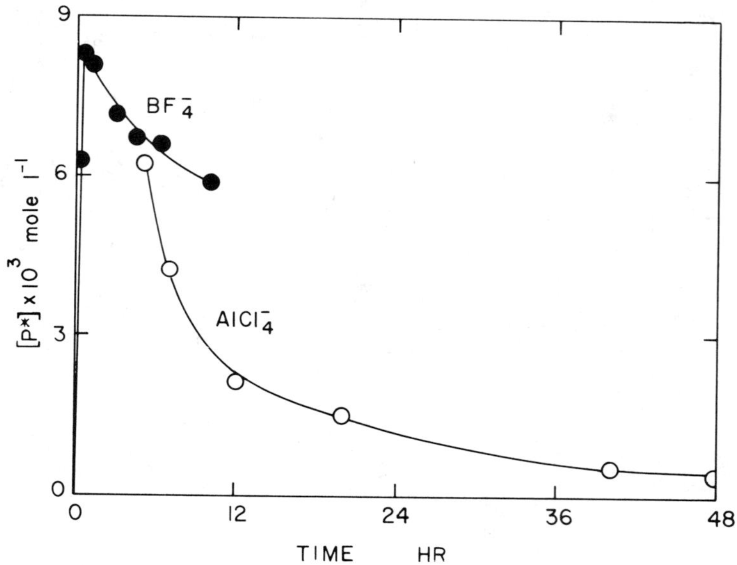

FIGURE 4.1 Variation of [P*] in polymerizations of THF initiated by $Et_3O^{\oplus}BF_4^{\ominus}$ (filled circles) and $Et_3O^{\oplus}AlCl_4^{\ominus}$ (open circles) in CH_2Cl_2 at 0°C.[17] (Reprinted from J. Macromol. Chem.–Chem., A-4, 873 (1970) by courtesy of Marcel Dekker Inc.)

from the final polymer or copolymer may lead to undesirable side reactions during storage.

The importance of both the cationic and anionic portions of the initiator was revealed by a study of various initiators for THF polymerization having different cations and different counterions. In this study, Yamashita *et al.*[21,22] concluded that the k_p value in CH_2Cl_2 at 0°C is almost independent of counterion when triethyloxonium ions are used as initiators. They found much slower apparent rates when the different cation initiators, acetyl hexafluoroantimonate and 2-methyl-1,3-dioxolenium perchlorate, were used. They explained the slower rates by decreased rates of initiation. The apparent k_p values can be increased to the usual value by the addition of

TABLE 4.1
Kinetic Parameters for THF Polymerization in the Absence of Esters

Initiator	Temp., °C	Solvent	E_p, kcal mole^{-1}	$10^{-8}A_p$, l mole^{-1} s^{-1}	10^2k_i, l mole^{-1} s^{-1}	10^3k_p, l mole^{-1} s^{-1}	10^2k_d, s^{-1}	10^5k_t, s^{-1}	Ref.
$Et_3O^+BF_4^-$	0	$(-CH_2Cl)_2$			1.3–1.7[a]	4.83			16
$Et_3O^+BF_4^-$	0	CH_2Cl_2			0.02[b]	3.7		1[b]	17
$Et_3O^+BF_4^-$	0	CH_2Cl_2				4.83			21
$Et_3O^+SbCl_6^-$	20	Bulk	15						25
$Et_3O^+PF_6^-$	0	CH_2Cl_2				4.19			21
$Et_3O^+SbF_6^-$	0	CH_2Cl_2				5.89			21
$Et_3O^+AlCl_4^-$	0	CH_2Cl_2			0.04			1000	17
$Et_3O^+ClO_4^-$	0	CH_2Cl_2				4.10			21
$CH_3CO^+SbF_6^-$	0	CH_3NO_2				1.47[c]			21
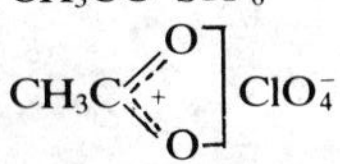	0	CH_3NO_2				1.65[c]			21

$BF_3 \cdot Et_2O$–ECH	0–40	Bulk	13.3	1.64				23
$BF_3 \cdot Et_2O$–ECH	20	Et_2O			16.6	4.67		23
$BF_3 \cdot$THF–ECH	0	Bulk			4.5			27
$BF_3 \cdot$THF–ECH	0	CH_2Cl_2	12	0.11	4.1			27, 28
$BF_3 \cdot$THF–ECH	20	Bulk			11.2 ± 2		11 ± 2[d]	29
$BF_3 \cdot$THF–PO	20	Bulk			32.3 ± 3		17 ± 3[d]	30
$SnCl_4 \cdot$THF–ECH	0	Bulk			6.2[e]			27
$AlEtCl_2 \cdot$THF–ECH	0	Bulk			7.8			27
$AlEt_3$–H_2O–ECH[f]	0	Bulk			6.6			33
$AlEt_3$–H_2O–PO[f]	0	Bulk			6.0			33
$AlEt_3$–H_2O–BPL[f]	0	Bulk			6.7			33

[a]Initial catalyst in range $1.53–6.1 \times 10^{-2}$; $[M_0] = 6.1$ M.
[b]Initial catalyst 9×10^{-3} M.
[c]Calculated assuming rapid initiation which was shown to be invalid.
[d]Called deactivation rate constant by investigators.[29,30]
[e]Average value; see Table 4.2.
[f]See Table 4.3.

epichlorhydrin (ECH) or trimethylene oxide, but only at the expense of some transfer early in the polymerization. Unfortunately, the number of growing oxonium ions was not experimentally determined in these experiments. The rate data are given in Table 4.1.

4.2.2 Triethyloxonium tetrafluoroborate

Vofsi and Tobolsky[16] reported the first careful kinetic study of THF polymerization. They used $Et_3O^+BF_4^-$ as initiator and carried out the polymerizations at 0°C in dichloroethane. By using ^{14}C-labeled catalyst, they were able to show that initiation was fast but not immediate. The rate of initiator disappearance was measured. The initial propagation rate was directly proportional to initial monomer concentration. The fit to the usual kinetic equation was good. The values obtained for k_i and k_p are given in Table 4.1. They did not determine a termination rate coefficient but realized that a complete description of the system would require one.

As stated above, Saegusa and Matsumoto[17] were able to estimate the importance of the termination term by using the [P*] method. The rate coefficients they found are given in Table 4.1.

Rosenberg *et al.*[23] generated their catalyst *in situ* from ECH and $BF_3{\cdot}Et_2O$. They carried out their polymerizations in bulk and in Et_2O solution. The experimental data are reported to fit the usual equation. They define $[I]_0$ in terms of the catalyst components charged assuming instantaneous and quantitative conversion to catalyst, but do not make any direct measurements of active centers. Their kinetic parameters are given in Table 4.1.

Before leaving $Et_3O^{\oplus}BF_4^{\ominus}$-initiated polymerizations of THF, one other study needs to be mentioned. Sangster and Worsfold[24] have found that, in THF polymerizations with $BF_4^{\ominus}$ counterions in CH_2Cl_2, the rate of polymerization of free ions is only seven times greater than that of ion pairs. Even this small difference is now open to question. (see the discussion of propagation rate constants in Section 4.2.6.)

4.2.3 Triethyloxonium hexachloroantimonate

Lyudvig *et al.*[25] studied the kinetics of bulk THF polymerizations initiated by $Et_3O^+SbCl_6^-$ at 20°C. In another study using this initiator, Stejny[26] examined the effect of simultaneous polymerization and crystallization of THF. He found that at his reaction temperatures, −25° and −45°C, at the onset of crystallization the polymerization rate was somewhat accelerated.

4.2.4 Lewis acid · THF · epoxide

In a kinetic study of bulk THF polymerizations at 0°C, Saegusa and Matsumoto[27] examined the effect of changing the Lewis acid in the system Lewis acid·THF·ECH. The phenoxy end-capping procedure was employed to determine [P*]. Kinetics were fitted to the equation with [P*], and k_p values were determined in the usual way. The pattern of [P*] variation during polymerization depended considerably on the Lewis-acid component. For the BF_3·ECH system [P*] increased slowly at first and then remained almost constant. In polymerizations initiated by $SnCl_4$·ECH, initiation was very rapid, and [P*] decreased very rapidly because of a termination reaction. Conversion to polymer was low. The course of the polymerization after initiation by the $AlEtCl_2$·ECH system was similar to that with $SnCl_4$·ECH, except that initiation was somewhat slower. In spite of these differences in initiation and overall conversion, there was not a wide difference in the k_p values obtained with the various initiators. The data are shown in Table 4.2.

Barzykina *et al.*[29,30] have made kinetic studies of THF polymerizations in bulk at 20°C using both the BF_3·ECH and BF_3·propylene oxide (BF_3·PO) systems. The rate constents that they found are given in Table 4.1. The rates were determined by dilatometry, and a separate determination of active centers was made. In both cases they observed a simultaneous deactivation process (reduction of number of active centers with time). Their studies led them to propose some of the termination reactions for the $BF_4^{\ominus}$ counterion summarized in Chapter 3.

TABLE 4.2
Effect of Lewis Acid on Propagation-rate Coefficient of Bulk-THF Polymerization at 0°C[27]

Lewis acid			
Type	M	$[ECH]_0$, M	$10^3 k_p$ $l\,mole^{-1}\,s^{-1}$
BF_3	0.011	0.010	4.5
BF_3	0.021	0.020	4.6
BF_3[a]	0.025	0.020	4.1
$SnCl_4$	0.028	0.029	5.8
$SnCl_4$	0.055	0.053	6.7
$AlEtCl_2$	0.054	0.051	7.8

[a]This polymerization was carried out in CH_2Cl_2 solution at $[M]_0 = 6.3$ M.

4.2.5 Triethylaluminum–water–promoter system

Saegusa *et al.*[31–33] carried out studies of the polymerization of THF initiated by the system $AlEt_3$–H_2O(2:1)–promoter. ECH, propylene oxide (PO) and β-propiolactone (BPL) were employed as promoters. The concentration of active sites was determined according to the phenoxyl end-capping method in

TABLE 4.3
Effect of Promoter on Propagation rate Coefficients of Bulk THF Polymerizations[a] at 0°C[33]

Promoter			
Type	M	$[AlEt_3]_0$[b] M	$10^3 k_p$, $l\,mol^{-1}\,s^{-1}$
ECH	0.02	0.18	6.6
PO[c]	0.03	0.15	6.0
BPL[d]	0.005	0.30	6.7

[a]Initiated by $AlEt_3$–H_2O (2:1) promoter.
[b]Amount of $AlEt_3$ used in preparation of catalyst.
[c]Propylene oxide.
[d]β-Propiolactone

studies after 1968, or calculated from yield and molecular weight before that date;[31] the former gave more reliable results. Kinetics could be fitted to the equation with [P*]. The kinetic constants observed are shown in Table 4.3. The k_p values for the three promoters are almost the same. They are compared with k_p values from other initiators in Table 4.1. The data show that the nature of the promoter does not affect the propagation reaction, although it does affect the initiation reaction and the overall rate. In the case of PO, a termination reaction was thought to be responsible for a slower rate of conversion.

4.2.6 Superacid esters

As discussed above, the kinetics of THF polymerization appear to be relatively simple. There is some disagreement about absolute values of the specific rate constants, but this can be attributed to experimental variables such as the use of counterions with variable stability, initiators with different efficiencies of generating active propagating species and/or different degrees of sophistication in experimental techniques. However, by and large, there has been reasonable agreement in the data obtained by various investigators.

The discovery of the superacid initiators changed this situation dramatically for several years. The reversible nature of the propagation with sufficiently stable complex anions continued to be accepted and the "living" character of the process was not questioned. However, the lack of dependence of k_p on counterion was challenged, and agreement on values of kinetic parameters vanished. Now, thanks to ingenious and elegant work by several groups of investigators, it is fair to say again that the kinetics of THF polymerization are well understood.

Major contributions were made by:

1) Smith and Hubin, who first postulated the existence of the macroion–macroester equilibrium;[34]

2) Saegusa, Kobayashi and coworkers, who, among other things, first examined the kinetics of THF polymerizations

initiated by a wide variety of superacid esters in a variety of solvents;[11,35]

3) Pruckmayr and Wu, who among other things introduced many of the NMR techniques;[36–38] and

4) Penczek, Matyjaszewski and coworkers, who have, it may be argued, made the most definitive kinetic studies to date, using selected superacid esters as well as initiators with a variety of complex anions.[13,39–48] Only this group has used experimentally measured equilibrium monomer concentrations, initiator concentrations comparable to those typically found in THF polymerizations and high-vacuum techniques with rigorously purified reagents throughout their studies. In addition, they have determined several thermodynamic equilibrium constants and the variation of propagation rate constants as a function of dielectric constant. It is beyond the scope of this chapter to reiterate all the arguments that led to such a careful study; but as the results and conclusions are presented below, it will be obvious that the information was needed, if valid kinetic rate constants were to be obtained and valid conclusions to be reached.

The chemical reactions As stated in Chapter 3, esters of strong acids are excellent initiators for the polymerization of THF. All the data presented so far in this chapter has assumed that the only equilibrium of interest involves THF and its tertiary oxonium ion. With esters of strong acids as initiators, a complication arises: in addition to the propagation and depropagation reactions involving oxonium ions, the equilibrium between the oxonium ion and an ester formed with the counterion must be taken into account. The chemical reactions that can occur are illustrated for $ROSO_2CF_3$ in Scheme 4.1. At each stage of the polymerization, the oxonium ion–counterion complex is in equilibrium with its corresponding ester.

The importance of NMR spectroscopy Many of the early analyses of THF polymerization kinetics with complex ion

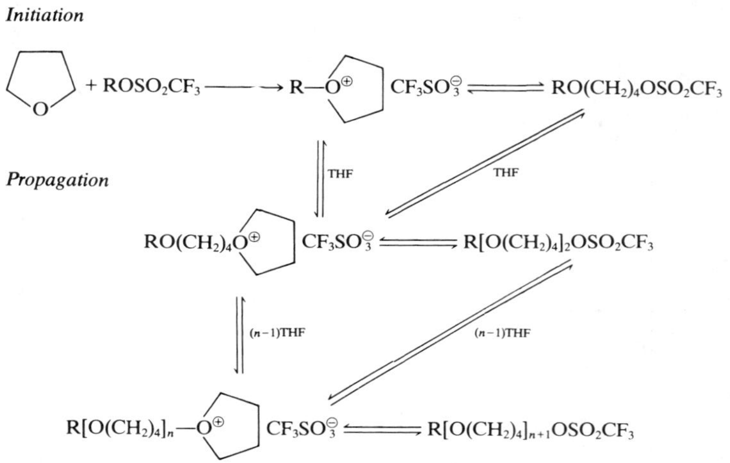

SCHEME 4.1 Chemical equilibria in THF polymerizations initiated with superacid esters.

initiators were carried out in bulk THF. A few were carried out in solution, especially in CH_2Cl_2. Kinetic analyses using superacid ester initiators have been carried out almost exclusively in solution. One of the reasons is that investigators learned very early in their studies that systems containing macroesters were readily analyzed by NMR spectroscopy: 100–300 MHz ^{1}H, ^{19}F and ^{13}C.[34–38] (^{19}F and ^{13}C NMR are preferred when they can be used, because of limitations of sensitivity and resolution in ^{1}H NMR.) The beauty of the system is that the peaks in the NMR spectra for initiator, oxonium ion and ester are well separated from each other, and the instantaneous concentrations of each can be determined easily. Typical chemical shifts are given in Table 4.4, and representative spectra can be found in Figures 4.2–4.4. The chemical shifts given for ^{1}H NMR are the ones currently believed to be correct. Some of the assignments were confused in the original paper dealing with use of NMR spectroscopy in these superacid-ester systems: the reader should beware.

TABLE 4.4
NMR Assignments for Polymerizations Initiated with Superacid Esters

Assignment[a]	^1H, PPM δ from TMS[b,c,d]	^{19}F, PPM δ from C_6H_6[d]	^{19}F, PPM δ from CF_3CO_2H[e]	^{19}F, PPM δ from CCl_3F[f]	^{13}C, PPM δ from TMS[g]
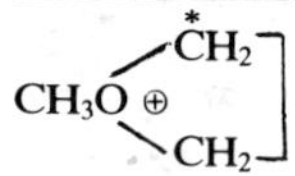					90.1
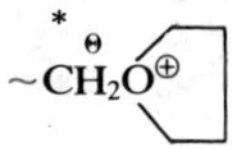	4.87(t), 4.9(m)				89.6
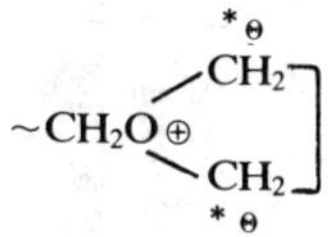	4.83_5(t)				87.5
$CH_3\overset{\theta}{CH_2}OSO_2CF_3$	4.75(qu)	87.8(s)	−2.52[i], −2.41[j]	−76.0[k], −75.8[l], −76.0[k,m]	
$\sim\overset{\theta}{CH_2}OClO_3$	4.64[h]				
$\sim\overset{\theta}{CH_2}OSO_2CF_3$	4.58(t)	88.1(s)	−2.80[i], −2.73[j]	−75.7[k,m]	
$CF_3SO_3^{\ominus}$				−79.0[k], −78.4[l], −78.7[l,m]	
$\sim\overset{*\theta}{CH_2}OSO_2F$				34.5[k,m]	79.1

$\sim\overset{*}{C}H_2O\overset{*}{C}H_2\sim$				71.1
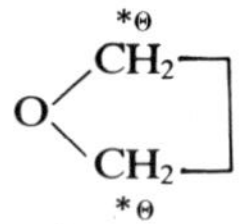	3.7(m)			68.2
$\overset{\ominus}{C}H_3OSO_2CF_3$	4.30(s)	88.6(s)		
$\overset{*}{C}H_3OSO_2F$				62.5
$\sim CH_2O\overset{*}{C}H_3$				58.4
$CH_3CH_2OSO_2F$			38.4[k], 35.0[l], 35.1[l,m]	
$FSO_3^{\ominus}$			38.8[k], 38.6[l], 38.0[l,m]	
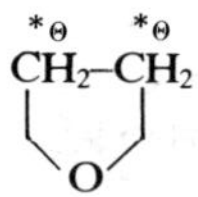	1.8(m)			26.2
$\sim OCH_2\overset{*\ominus}{C}H_2\overset{*\ominus}{C}H_2CH_2O\sim$	1.61(t)			27.4
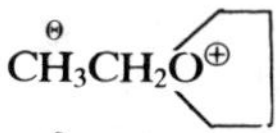	1.6(t)			
$\overset{\ominus}{C}H_3CH_2OSO_2CF_3$	1.5(t), 1.45(t)	87.8(s)	−76.0[k], −75.8[l], −76.0[k,m]	

[a–m]See p. 90 for footnotes to table.

[a]Carbon of interest in ^{13}C NMR is marked with an asterisk. Protons of interest in ^{1}H NMR are marked with the symbol "Θ", ^{1}H NMR chemical shifts do not depend on the concentration of the species being measured. The fluorine assignments are obvious; ^{19}F NMR chemical shifts change slightly with concentration.

[b]Measured with Varian instrument, 300 MHz.

[c]Abbreviations: s, singlet, t, triplet, qu, quartet, m, multiplet.

[d]Penczek, Matyjaszewski, *et al.*[39,41,44] The ^{19}F data are from Buyle *et al.*[43]

[e]Saegusa and Kobayashi.[11]

[f]Pruckmayr and Wu.[36]

[g]Wu and Pruckmayr.[37]

[h]Matyjaszewski, *et al.*[47] The value given is for a 250 MHz FT ^{1}H NMR spectrometer. The oxonium ion protons had about the same assignments as in the 300 MHz instrument.

[i]In CCl_4. The chemical shifts were solvent dependent.

[j]In CH_2Cl_2. The chemical shifts were solvent dependent.

[k]In CCl_4. The chemical shifts were solvent dependent.

[l]In CH_3NO_2. The chemical shifts were solvent dependent.

[m]With THF also. The chemical shifts were solvent dependent.

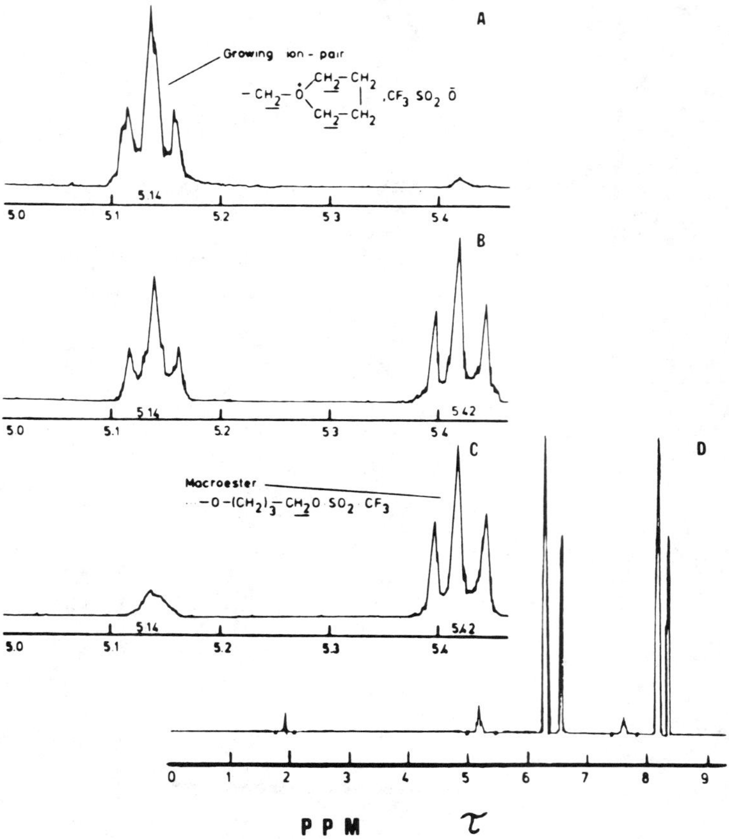

FIGURE 4.2 300-MHz ^{1}H-NMR spectra of living PTHF at 17°C: (A) in CD_3NO_2, $[M]_0 = 8.0$ M, $[M]_e = 4.65$ M, $[I]_0 = 0.05$ M; (B) in CH_2Cl_2, $[M]_0 = 8.0$ M, $[M]_e = 4.9$ M, $[I]_0 = 0.05$ M; (C) in CCl_4, $[M]_0 = 8.0$ M, $[M]_e = 3.4$ M, $[I]_0 = 0.1$ M; (D) in CD_3NO_2, $[M]_0 = 8.0$ M, $[M]_e = 4.65$ M, $[I']_0 = 0.13$ M. Here I is ethyl trifluoromethanesulfonate and the counterion for A, B and C is $CF_3SO_2O^{\ominus}$; I′ is 1,3-dioxolan-2-ylium hexafluoroantimonate and the counterion is $SbF_6^{\ominus}$. Only the region from 5.0 to 5.5 τ is shown in spectra A, B and C. Spectrum D is the complete spectrum with the regions from 1.70 to 2.10, from 5.0 to 5.5 and from 7.4 to 7.8 τ at 10-fold increased amplitude.[40]

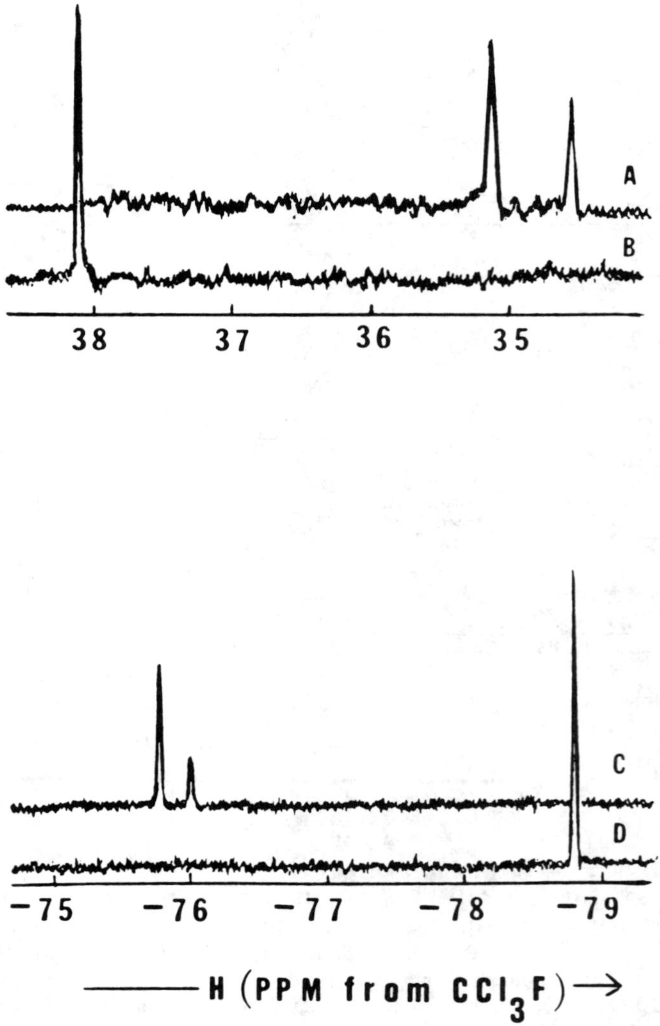

FIGURE 4.3 56.4-MHz ^{19}F-NMR spectra of THF–superacid-ester mixtures (10:1) after 60 min polymerization time: (A) in CCl_4 with $EtOSO_2F$; (B) in CH_3NO_2 with $EtOSO_2F$; (C) in CCl_4 with $EtOSO_2CF_3$; (D) in CH_3NO_2 with $EtOSO_2CF_3$. The solvent concentration was 66.6 wt% in all cases.[37]

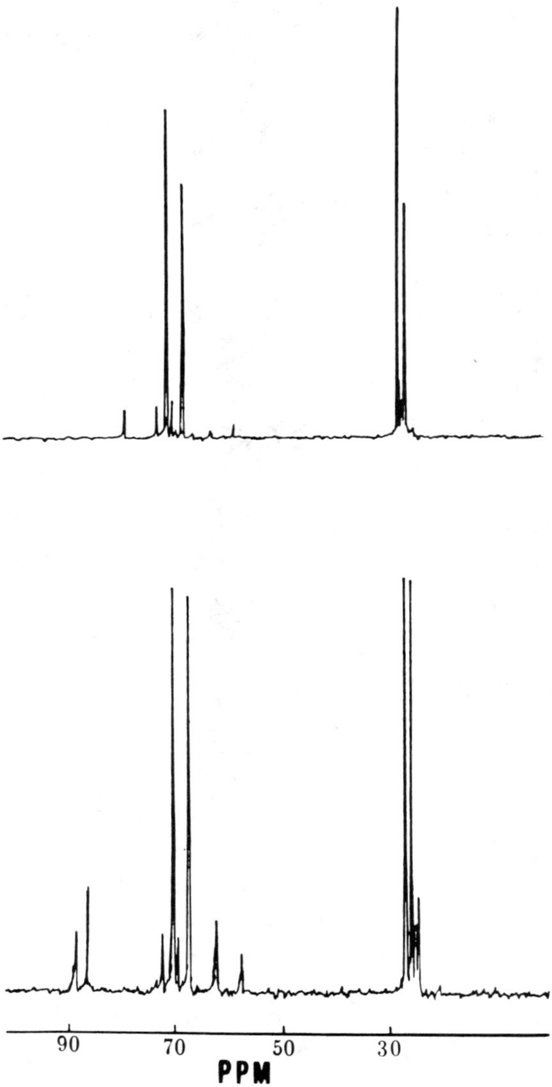

FIGURE 4.4 22.63 MHz ^{13}C NMR spectra of THF–CH_3OSO_2F (6:1) mixtures after 20 min polymerization time: (A) in CH_3NO_2 (64%); (B) in CCl_4 (64%).[38]

TABLE 4.5
Kinetic and Thermodynamic Constants in THF Polymerizations

Counterion	Initiator	$[I]_0$ M	Solvent	$\sim\epsilon$ of solvent[a]	$[M]_0$, M	ϵ of mixture[a,b]	X_i[c]	T, °C	k_i or $k_i^E \times 10^5$, $l\ mol^{-1}\ s^{-1}$	$k_p^{app} \times 10^3$, $l\ mol^{-1}\ s^{-1}$	$k_p^i \times 10^3$ $l\ mol^{-1}\ s^{-1}$	$-\Delta H_p^{\neq}$ kJ mol^{-1}	$-\Delta S_p^{\neq}$, J mol^{-1} deg^{-1}	Ref.
$SO_3CF_3^{\ominus}$	d	0.053–0.067	CCl_4	2.24	8.0	5.9	0.045	25	6	1.6	44	37.7	138	39, 41, 45
	e	0.10	CCl_4	2.24	8.0	5.9	0.035	25	25	16	40			41
	d	1.05	CCl_4	2.24	5.3	4.5	0.45	0	0.80	0.84	2.2	66.9	50	11
	d	1.05	CCl_4	2.24	5.3	4.5	0.47	25	3.9	3.0	5.7			11
	d	~0.06	CCl_4	2.24	5.0	4.6	0.02	25	3.35	0.95	51.2			47
	d	~0.06	CCl_4	2.24	6.0	5.0	0.03	25	4.4	1.4	46.7			47
	d	~0.06	CCl_4	2.24	7.0	5.5	0.037	25	5.6	1.6	43.3	35.1	151	47
	d	~0.06	CCl_4	2.24	8.0	5.9	0.045	25	6.7	1.8	40.7			47
	d	~0.06	CCl_4	2.24	9.0	6.3	0.06	25	7.9	2.3	38.6			47
	d	~0.06	CCl_4	2.24	10.0	6.7	0.08	25	9.1	3.0	36.9			47
	d	1.05	C_6H_6	2.28	5.3	4.5	0.75	0	1.1	1.7	2.7			11
	d	1.05	$CHCl_3$	4.81	5.3	6.0	0.85	0	1.1	1.8	2.2			11
	d	1.05	CH_2Cl_2	8.93	5.3	8.4	0.89	0	1.2	1.7	2.0	48.5	121	11
	d	0.05	CH_2Cl_2	8.93	8.0	8.1	0.23	25			39.2			46
	d	1.05	$C_6H_5NO_2$	34.8	5.3	23.4	>0.98	0	3.1	4.4	4.4			11
	d	0.036	CH_3NO_2	41.5	7.0	25.7	~1.0	0	2050	20.5	20.5			45
	d	0.05	CH_3NO_2	41.5	8.0	22.8	~1.0	0	2100	21	21	50.2	100	45

$SO_3F^{\ominus}$	f	—	CCl_4	2.24	8.0	5.9	—	25	9.6	—	—			41
	f	0.01	CH_3NO_3	41.5	7.0	25.7	0.98	25	2200	22.1	22.1			45
$A^{\ominus}$	g	0.02	CCl_4	2.24	6.0	5.0	1.0	25	4700	47	47			41
$AsF_6^{\ominus}$	g	—	CCl_4	2.24	8.0	5.9	1.0	25	4600	46	46			39
	g	0.02	CH_2Cl_2	8.93	6.8	8.3	1.0	25	6300	63	63			46
	g	0.001	CH_3NO_3	41.5	7.0	25.7	1.0	25	2000	20.2	20.2			45
$SbF_6^{\ominus}$	g	~0.02	CCl_4	2.24	8.0	5.9	1.0	25	4600	46	46			41
	g	0.0016	CH_2Cl_2	8.93	7.1	8.2	1.0	25	6300	63	63			46
	g	~0.02	CH_3NO_3	41.5	7.0	25.7	1.0	25	2100	21	21			45
	g	0.13	CH_3NO_3	41.5	8.0	22.8	0.92	25	4000	40	40			40
	g		THF	7.58	12.3	7.58	1.0	25	3400	34	34	58.6	73.2	41, 45
$SbCl_6^{\ominus}$	g		CCl_4	2.24	8.0	5.9	1.0	25	4600	46	46			39
$PF_6^{\ominus}$	g		CCl_4	2.24	8.0	5.9	1.0	25	4600	46	46			39
$BF_4^{\ominus}$	g		CCl_4	2.24	8.0	5.9	1.0	25	4600	46	46			39

[a]Dielectric constant taken from Saegusa and Kobayashi[11] or, in the case of nitromethane, from Dean.[49]

[b]Calculated by assuming molar additivity of the dielectric constant values of solvent and THF. Since Saegusa's values for ϵ were used, these numbers differ somewhat from those stated by Penczek. These values were not corrected for differences in temperature of polymerization and are all based on 0°C, on the assumption that Saegusa's values are for 0°C.

[c]$X_{\bar{I}}$ is the mole fraction of macroions of all kinds at equilibrium. Saegusa *et al.*, use the term $[O^+]$ fraction and Penczek *et al.* use β. In some work by Penczek *et al.*, α is used as the mole fraction of macroesters; $\alpha + \beta = 1$.

[d]$EtOSO_2CF_3$.

[e]$MeOSO_2CF_3$.

[f]$MeOSO_2F$.

[g]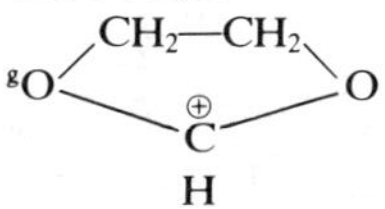 A^- salts, where A^- is a general anion. Values for specific anions are given also.

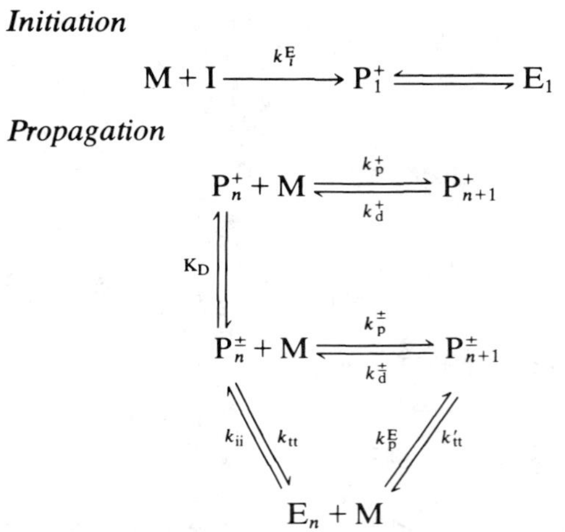

SCHEME 4.2 Following primarily Penczek and coworkers, the symbols used are: M, monomer; I, initiator; P_n^+, macroion paired with anion and containing n monomer units (Saegusa and coworkers do not distinguish between P_n^+ and $P_n^{\pm}$, but use O^+ for all ions); E_n, macroester containing n monomer units (Saegusa and coworkers use E for macroesters); k_i^E, rate constant of initiation by esters of superacids; k_p^+, rate constant of propagation for unpaired macroions; k_d^+, rate constant of depropagation for unpaired macroions; $k_p^{\pm}$, rate constant of propagation for macroion pairs; $k_d^{\pm}$, rate constant of depropagation for macroion pairs; k_p^E, rate constant of propagation for macroesters ($\simeq k_i$ for superacid esters from the work of Penczek and coworkers); k_{ii}, rate constant for internal ionization of macroesters to macroion pairs; k_{tt}, k'_{tt}, rate constant for conversion of macroion pairs into macroesters (as far as propagation is concerned the macroesters are in a state of "temporary termination"); K_D, equilibrium dissociation constant for macroion pairs into macroions. In addition, the complete description of the system also requires the determination of: $K_e = k_{ii}/k_{tt}$, the equilibrium constant for interconversion of macroesters into macroion pairs; k_p^{app}, the apparent rate constant for propagation as observed by, say, dilatometry and calculated according to the usual equation assuming $[I]_0$ is equal to the instantaneous concentration of active chain ends, that is $k_p^{app} = k_p^{\bar{I}} X_{\bar{I}} + k_p^E X_E$, where $\bar{I}$ denotes macroions of all kinds (paired and unpaired) and $k_p^{\bar{I}}$ is the rate constant of propagation for macroions of all kinds (in general, $k_p^{\pm} = k_p^+ = k_p^{\bar{I}}$, according to Penczek and coworkers; (note that Saegusa and coworkers use k_p^{ap} for k_p^{app}) while $X_{\bar{I}}$ and X_E are the mole fractions of macroions of all kinds (P_n^+ and $P_n^{\pm}$) and of macroesters (E_n), respectively (Saegusa and coworkers use $[O^+]$ for $X_{\bar{I}}$): $X_E = \alpha$, the degree of dissociation $= [E_n]/([E_n] + [P_n^{\pm}] + [P_n^+]) = [E_n]/[I]_0$; $X_{\bar{I}} = \beta = ([P_n^{\pm}] + [P_n^+])/([E_n] + [P_n^{\pm}] + [P_n^+])$.

It was immediately apparent from the NMR studies that, as shown in column $X_{\bar{I}}$ of Table 4.5, the ion–ester equilibria could be shifted almost completely to the ion or to the ester by proper choice of solvent. Thus, in CCl_4, only 2% of the growing species in a polymerization with $[THF]_0 = 5$ M and a trifluoromethanesulfonate ester initiator are macroions ($X_{\bar{I}}$ = mole fraction of macroions = 0.02 ≡ 2%). Under the same conditions in CH_3NO_2, 98% of the growing species are macroions. Thus, by judicious choice of solvent, it is possible to determine many kinetic constants that were previously inaccessible.

The kinetic constants and some thermodynamic equilibrium constants The kinetic parameters that need to be considered are shown in Scheme 4.2. In addition to rate constants for initiation, propagation, depropagation and ester–ion equilibria, the complete kinetic analysis also requires the determination of the equilibrium constants for dissociation of ion pairs into ions, K_D, and for macroions into macroesters, $K_e = k_{ii}/k_{tt}$. As stated, it is possible to determine all these constants. Most of the values are summarized in Table 4.5. Before discussing the significance of the results, several apparent anomalies that needed to be resolved will be briefly discussed. It is worthwhile to consider these problems, because they demonstrate once again the great care that is required and the many factors that must be considered in determining valid kinetic constants for THF polymerizations.

The macroion–macroester equilibrium The initial controversy over the assignments of chemical shifts in ^{1}H NMR, coupled with the problems in sensitivity and resolution in ^{1}H NMR, caused a brief period of doubt about the existence of the equilibrium between macroesters and macroions. With the aid of ^{19}F NMR, investigators were able to establish unequivocally the validity of the existence of macroesters (see data in Table 4.4).

The equilibrium monomer concentration, $[THF]_e = [M]_e$ When solutions of THF in various solvents were studied, it turned out that they do not behave ideally.[50,51] This is important in studies of the kinetics of THF polymerizations because, as a result, $[M]_e$ varies not only with temperature, as already known, but also with solvent. Figure 4.5 shows the variation of $[M]_e$ as a function of temperature and of initial monomer concentration, $[THF]_0$, in CH_2Cl_2 solutions. Figure 4.6 shows the variation of $[M]_e$ at 25°C as a function of initial monomer concentration in a variety of solvents. The complete thermodynamic analysis of such systems has been reported by Ivin and Leonard,[13] who derived the expression

$$\Delta G^0_{lc}/RT = 1 + \ln \phi_1 + \chi_{12}(\phi_2 - \phi_1) + (\chi_{13} - \chi_{23} V_1/V_3)\phi_3,$$

where ΔG^0_{lc} is the free energy for the reaction M(liquid)→

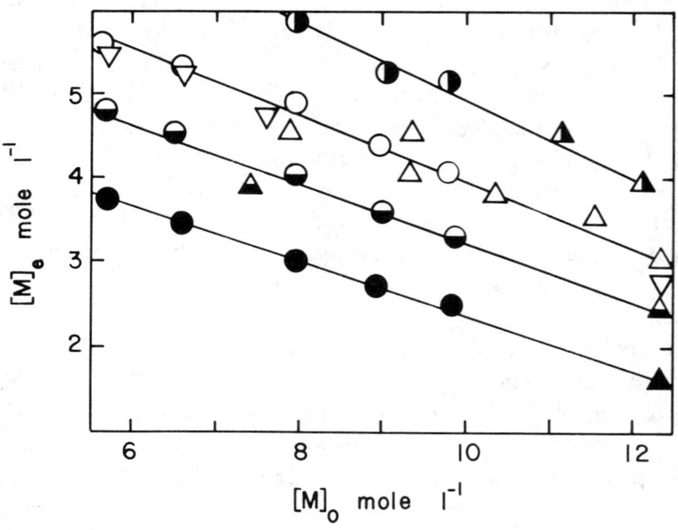

FIGURE 4.5 Effect of $[M]_0$ on $[M]_e$ in CH_2Cl_2 solvent and at different temperatures: 0°C (▲, ●); 15°C (▲, ◒); 25°C (△, ○); 35°C (▲, ◑). Data in circles were obtained by NMR; data in triangles were obtained by gravimetry;[46] data at 25°C (▽) were taken from Dreyfuss and Kennedy.[52]

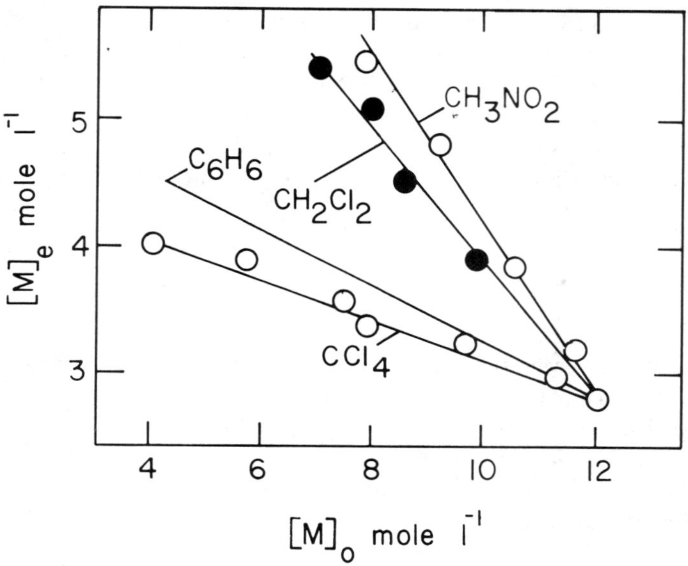

FIGURE 4.6 Effect of $[M]_0$ on $[M]_e$ at 25°C in different solvents.[13]

$(1/n)M_n$ (amorphous); ϕ_1, ϕ_2, ϕ_3 are mole fractions for monomer, polymer and solvent, respectively; V_1 and V_3 are corresponding volume fractions; χ_{12}, χ_{13} and χ_{23} are interaction parameters for monomer–polymer, monomer–solvent and polymer–solvent, respectively.[13]

Initiation Penczek, Matyjaszewski and coworkers used two main groups of initiators for their kinetic studies. One group consisted of superacid esters, primarily $EtOSO_2CF_3$ and $MeOSO_2CF_3$; a few experiments with $MeOSO_2F$ were also carried out. They found that the methylation and ethylation required for initiation are quantitative with no parallel or consecutive reactions, provided that $[THF]_0$ is low enough and the conversion to polymer is less than 25%. The rate of initiation is slow. It depends on the polarity of the medium and k_i is generally about 1000 times less than k_p. This slow rate of initiation results in nonlinear plots, when the data are plotted in the ordinary way. An example is shown in Figure 4.7. Penczek, Matyjaszewski and coworkers have derived the following equation to describe THF polymerizations under these conditions:

$$\ln \{\mathrm{d} \ln ([\mathrm{M}]_t - [\mathrm{M}]_e)/[\mathrm{I}]_0 \,\mathrm{d}t + k_p^{\mathrm{app}}\} = \ln k_p^{\mathrm{app}} - k_i \int_0^t [\mathrm{M}]_t \,\mathrm{d}t.$$

Following this equation, k_i can be determined from the slope of the plot of the left-hand side of the equation against $\int_0^t [\mathrm{M}]_t \,\mathrm{d}t$ determined graphically. Another method used to determine k_i was by ^{1}H NMR spectroscopy.

The other group of initiators used by Penczek, Matyjaszewski and coworkers consisted of 1,3-dioxolan-2-ylium

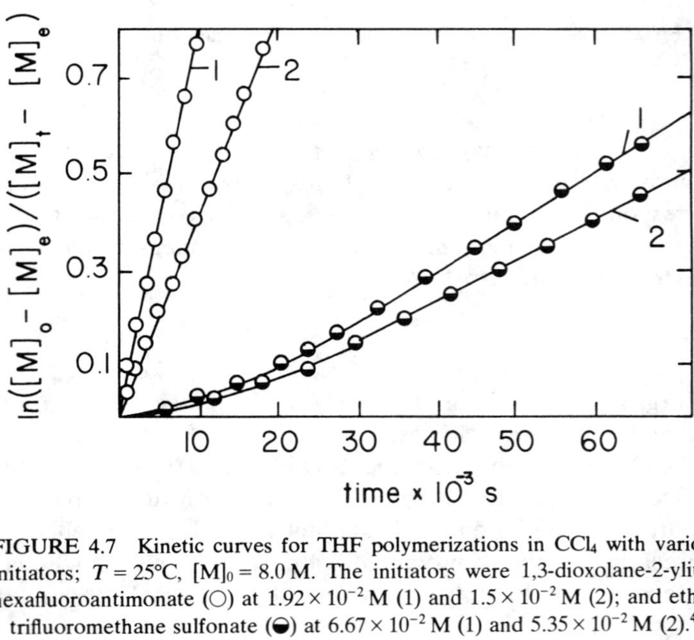

FIGURE 4.7 Kinetic curves for THF polymerizations in CCl_4 with various initiators; $T = 25°C$, $[M]_0 = 8.0$ M. The initiators were 1,3-dioxolane-2-ylium hexafluoroantimonate (○) at 1.92×10^{-2} M (1) and 1.5×10^{-2} M (2); and ethyl-trifluoromethane sulfonate (◒) at 6.67×10^{-2} M (1) and 5.35×10^{-2} M (2).[41]

salts, $\overline{OCH_2CH_2OCH}^{\oplus}A^{\ominus}$, with various anions, $A^{\ominus}$, including $SbF_6^{\ominus}$, $PF_6^{\ominus}$, $SbCl_6^{\ominus}$ and $BF_4^{\ominus}$. They report that these salts are very convenient initiators for studying the polymerization of THF. Initiation is not accompanied by side reactions, and the rate of initiation can be studied by ^1H NMR spectroscopy by following the disappearance of either the CH proton at 8.1 δ or the CH_2 protons at (9.12–9.45) δ.[41] Only macroions form in the presence of these anions. Initiation is not fast, however. As can be seen from Table 4.5, k_i is equal to k_p^I, the rate of propagation of macroions. Since k_i and k_p^I are equal, kinetic curves plotted in the usual way are linear (cf. Figure 4.4). Ultimately, initiation was complete and efficient. The number of moles of growing centers was equal to the number of initiator molecules used. Molecular weights obtained by high-speed osmometry agreed well with those calculated from $[I]_0$ and the number of moles of THF polymerized.

Propagation k_p^{app} The easiest of the propagation rate constants to determine is k_p^{app}, the apparent rate of propagation. Most often k_p^{app} is determined dilatometrically.

When esters of superacids are used as initiators, k_p^{app} can also be determined from the intercept of the plot described above for the determination of k_i. Penczek, Matyjaszewski and coworkers have checked their values of k_p^{app} in two ways. First, since they found that initiation with the anhydride of trifluoromethanesulfonic acid is fast and that no acceleration of the polymerization is observed, they determined k_p^{app} using the anhydride. They obtained a value of k_p^{app} in good agreement with that determined by the other methods.

(A few words of caution need to be considered by any reader planning to use the anhydride to determine k_p^{app}. Allowance must be made for the fact that the anhydride leads to a polymer capable of growing on both ends, whereas ester initiation produces a polymer growing at one end only. Additional complications can also arise if the relative concentration of anhydride is increased beyond about 2 mol% of the THF concentration. Smith, Schultz and Newmark[53] reported that comparatively stable $CF_2SO_3(CH_2)_4O_3SCF_3$ and THF-trimer bisoxonium salt can form and that, once formed, these species initiate THF polymerization slowly.)

Second, Penczek, Matyjaszewski and coworkers observed that, if they depolymerized a polymerization initiated with $EtOSO_2CF_3$ after a constant k_p^{app} was attained, and then repolymerized it, the same k_p^{app} was obtained.

Since esters do not form with complex ions, $k_p^{app} = k_p^{\bar{I}}$ is the only propagation rate constant observed. It can be determined dilatometrically or by NMR. (Here $k_p^{\bar{I}}$ is the rate constant of propagation of macroions of all kinds.)

k_p^+, $k_p^{\pm}$, $k_p^{\bar{I}}$ The values of k_p^+ and $k_p^{\pm}$ were determined in polymerizations of THF initiated by 1,3-dioxolan-2-ylium cations with $AsF_6^{\ominus}$ and $SbF_6^{\ominus}$ anions. The polymerizations were carried out in CH_3NO_2 or CH_2Cl_2 solvents, or in mixtures of the two. Some experiments using CCl_4 were also made. In this way, a range of dielectric constants from five to 22 and a range of degrees of dissociation, α, of macroion pairs to macroions from 0.03 to 0.70 were achieved. Average $k_p^{app} = k_p^{\bar{I}} = \overline{k_p}$ values were calculated from the slopes of the usual kinetic plots derived from duplicate experiments. Here $\overline{k_p}$ is the average k_p^{app} determined from several independent experiments. The rate constants were found to be independent of counterion. As can be seen from Table 4.5, this independence of counterion was confirmed with other counterions, including those derived from superacids. (In CH_3NO_2 about 98% of the growing centers are present as macroions.) Penczek, Matyjaszewski and coworkers suggest that the rate of propagation is independent of the anion structure and of the state of aggregation of the ions, because the growing species are highly solvated or separated by THF molecules.

However, $k_p^{\bar{I}}$ was strongly dependent on the medium. This dependence on α and dielectric constant is shown in Figure 4.8 at 25°C and $[THF]_0 = 7.0$ M. The values of α used were calculated from the dissociation constant K_D using the equation

$$K_D = (|\tilde{I}|_0 \cdot \alpha^2)/(1 - \alpha).$$

The value of K_D itself was measured conductimetrically and

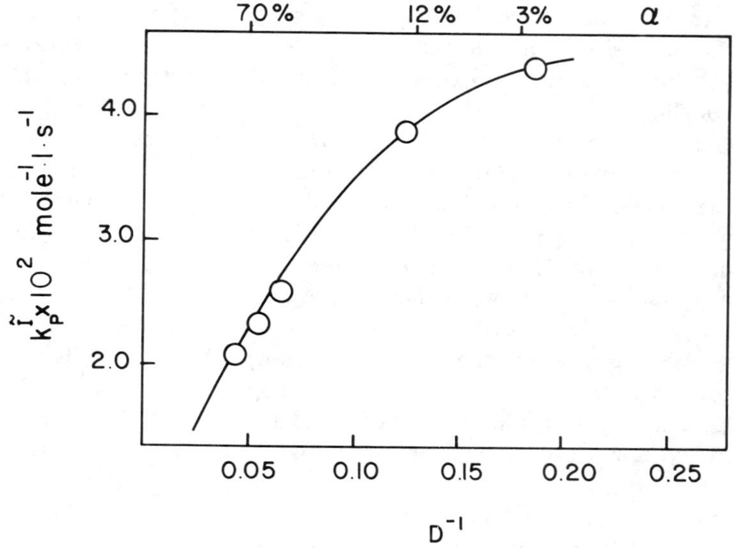

FIGURE 4.8 Effect of dielectric constant at 25°C and $[M]_0 = 7.0$ M on $k_p^{\tilde{I}}$.[46]

evaluated according to the Fuoss–Kraus method from plots of specific conductivity versus the total concentration of ions, $|\tilde{I}|_0$.[45] Typical values of K_D were 2×10^{-3} M at 25°C and $[THF_0] = 7.0$ M in CH_3NO_2 and 1.5×10^{-5} M at 25°C and $[THF]_0 = 7.0$ M in CH_2Cl_2.

Since $\bar{k}_p$ did not change very much with α or the dielectric constant, k_p^+ and $k_p^\pm$ could not be determined from the usual Szwarc plot.[54,55] Therefore, Penczek, Matyjaszewski and coworkers elected to treat the data accordingly to the equation:

$$\bar{k}_p = \alpha k_p^+ + (1-\alpha)k_p^\pm,$$

which when rearranged gives

$$\bar{k}_p/\alpha = (k_p^+ - k_p^\pm) + k_p^\pm/\alpha.$$

By plotting $\bar{k}_p/\alpha$ versus α^{-1}, $(k_p^+ - k_p^\pm)$ was obtained from the intercept and $k_p^{\mp}$ from the slope of the line that resulted from the plot. The measured values of $\bar{k}_p$, k_p^{+} and $k_p^{\mp}$ were found to be the same within experimental error, and these workers concluded that $k_p^{+} = k_p^{\pm} = k_p^{app} = k_p^{\bar{I}}$.

This conclusion that the reactivity of macroions and of macroion pairs in THF polymerizations is identical differs from the conclusion reached earlier by Worsfold and coworkers,[24,56] which was that macroions are 7–20 times more reactive than macroion pairs. Penczek *et al.* defend their results by pointing out that the earlier work did not use the correct $[M]_e$ and that the values were determined at low conversions (below 5%), where initiation was assumed to be complete but actually was probably far from complete. The important point is the agreement of both groups of workers that in THF polymerizations, especially compared to anionic polymerizations, there is little, if any, difference between the reactivity of macroion pairs and macroions.

k_p^E Penczek, Matyjaszewski and coworkers have further demonstrated that $k_i^E \simeq k_p^E$, i.e., that the rate of initiation with esters of superacids is approximately equal to the rate of propagation of macroesters.[47] This means that k_p^E is 100 to 1,000 times smaller than $k_p^{\bar{I}}$. Thus, even though macroesters are present in very large excess in systems like those containing CCl_4, the polymerization rate is dominated by the rate constant for macroions.[47] In CCl_4 and CH_2Cl_2, k_p^{app} is much less for polymerizations with $CF_3SO_3^-$ and FSO_3^- anions than it is for systems initiated with a comparable amount of an initiator that results in a complex ion as counterion. The lower apparent rate is due to the high proportion of macroesters, which can be assumed to be a "sleeping" species.[11]

Dependence of $k_p^{\bar{I}}$ on temperature It has been demonstrated that, even though esters may be present in the system, the apparent rate of polymerization is dominated by $k_p^{\bar{I}}$. Consequently, it is sufficient to determine the effect of temperature on $k_p^{\bar{I}}$ if knowledge of the temperature dependence of the rate of THF polymerization is desired. An example of this variation is shown in Figure 4.9, where the dependence of $k_p^{\bar{I}}$ on reciprocal temperature in CH_2Cl_2 solvent and $[THF]_0 = 7$ M is

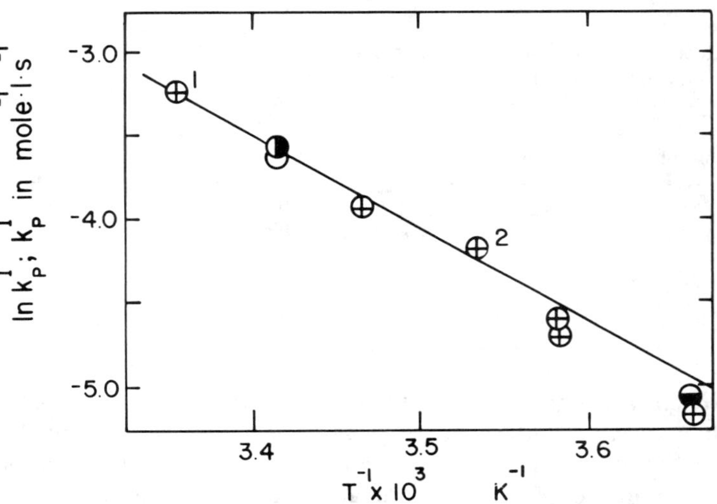

FIGURE 4.9 Effect of temperature on $\ln k_p^{\bar{I}}$ in CH_2Cl_2 solvent with $[M]_0 =$ 7 M. The counterions are $SbF_6^{\ominus}$(⊕), $AsF_6^{\ominus}$(◑), $BF_4^{\ominus}$(◒), $PF_6^{\ominus}$(○). The label 1 denotes an average from 15 measurements, and the label 2 an average from 10 measurements.[46]

given. Once the temperature dependences of $[M]_e$ and $k_p^{\bar{I}}$ are known, evaluation of $\Delta H_p^{\neq}$ and $\Delta S_p^{\neq}$ follow as before. Some values are given in Table 4.5. In the more polar solvent, CH_3NO_2, both $\Delta H_p^{\neq}$ and $\Delta S_p^{\neq}$ are higher. These higher values are unexpected, and it appears that solvation of the ground state by THF monomer is more important than the dielectric constant of the medium.[45]

The kinetics and thermodynamics of the interconversion of macroesters and macroion pairs have also been determined.[43]

K_D and K_E A method for obtaining K_D has already been described (p. 102) The value of $K_E = k_{ii}/k_{tt}$ (rate constant for internal ionization/rate constant for temporary termination) has been determined by ^1H (300 MHz) and ^{19}F NMR methods.[43] Some values of k_{ii}, k_{tt} and K_E are summarized in Table 4.6. It seems that K_E is determined mostly by the rate of macroion-pair collapse into macroester.

TABLE 4-6
Kinetic and Thermodynamic Parameters for Interconversion of Macroesters and Macroion Pairs in THF Polymerization Initiated with Triflate Ester (43)

Solvent	ΔH_e kJ mol^{-1}	ΔS_e J mol^{-10} K^{-1}	K_e [a]	k_{ii} [a] s^{-1} × 10^2	k_{tt} [a] s^{-1} × 10^2
CCl_4	−25.1 ± 1.3	−105 ± 4	0.07	0.8	12.1
CH_2Cl_2	−23.4 ± 0.8	−82.4 ± 3	0.58	1.9	3.3

[a]at 25°C.

Concurrent influences of temperature and polarity of medium on rate constants Matyjaszewski and Zielinski[57] have derived a mathematical equation containing four constants to describe the influence of temperature and dielectric constant of the medium on the rate and equilibrium constants in THF polymerizations. Since the dielectric constant of a medium varies with temperature, the equation gives the dependence on both parameters simultaneously. Values of the constants were determined.

Macroesters and macroions with $ClO_4^{\ominus}$ counterions It has long been suspected that macroesters should form in the presence of perchlorate ions, but experimental verification has been obtained only recently with the aid of FT–^{1}H NMR at 250 MHz.[48] Some results of the study are given in Table 4.7. The acid strengths of the three acids listed are: $HSO_3CF_3 >$ $HSO_3F > HClO_4$. It is evident that stronger acids lead to higher K_e values and greater proportions of macroions. The reason that esters have not been confirmed previously with $ClO_4^{\ominus}$ anions is probably that the chemical shift, shown in Table 4.4, is even a little closer to the corresponding ion than it is for $CF_3SO_3^{\ominus}$ and $FSO_3^{\ominus}$ anions, for which convincing data on the existence of esters have been obtained only since 250–300 MHz ^{1}H NMR and ^{19}F NMR and ^{13}C NMR spectroscopic methods have been available.

TABLE 4.7
Equilibrium constants, K_e,[a] between[48] Macroions and Macroesters[b]

Counterion	K_e CH_3NO_2	K_e CH_2Cl_2	Acid strength[c]	Nucleophilicity of anion[c]	Relative amount of ester[c]
$CF_3SO_3^{\ominus}$	42	0.58	1	3	3
$FSO_3^{\ominus}$	20	0.27	2	2	2
$ClO_4^{\ominus}$	9.5	0.09	3	1	1

[a] $K_e = [P^+] + [P^{\pm}]/[I]_0 = [P^+] + [P^{\pm}]/([P^+] + [P^{\pm}] + [E])$.

[b] $[THF]_0 = 8.0$ M; $T = 20°C$; $[I]_0$ = variable.

[c] The lowest number indicates relatively the highest value. Thus, the relative amount of ester in the presence of a given anion is: $ClO_4^- > FSO_3^- > CF_3SO_3^-$.

Concluding remarks and summary The important points arising out of the foregoing are as follows.

1) With complex anions, the kinetics of THF polymerization in the bulk are comparatively simple. The propagating species is the THF tertiary oxonium ion and the rate of propagation is independent of the accompanying anion. Since the polymerization of THF has a relatively low ceiling temperature (about 83°C) even in the bulk, the *equilibrium monomer concentration*, $[M]_e$ must be included in kinetic expressions. The data fit the expression

$$k_p t = \frac{1}{[I]_0} \ln \left\{ \frac{[M]_0 - [M]_e}{[M]_t - [M]_e} \right\}.$$

2) With complex anions, the kinetics of THF polymerization in solution are similar to those in the bulk, except that THF solutions do not behave ideally and the equilibrium monomer concentration must be determined under the experimental conditions.

3) The rates of propagation of THF macroions and macroion pairs are equal or nearly equal.

4) With anions that can form esters with the growing tertiary oxonium ion, the kinetics of propagation of THF polymerizations are dominated by the rate of propagation of the macroions. The proportion of macroions compared to macroesters for any given anion varies with the solvent–monomer mixture and must be determined before a kinetic analysis can be made. For all practical purposes, slower rates will be observed whenever macroesters form, even though the total number of growing centers is the same, because the esters can be considered to be in a state of "temporary termination"[43] or "sleeping".[11] Polymerization of the macroester occurs, but at a much slower rate than the polymerization of the macroion.

5) With anions that are known to form macroesters, the proportion of macroester formed decreases as the strength of the acid of the anion increases. Thus the proportion of

macroester is in the order: $ClO_4^{\ominus} > FSO_3^{\ominus} > SO_3CF_3^{\ominus}$; and the observed rate of polymerization at a given molar initiator concentration, monomer concentration and temperature is in the order: $SO_3CF_3^{\ominus} > FSO_3^{\ominus} > ClO_4^{\ominus}$.

6) THF polymerizations with many ions are free from termination and transfer to monomer or solvent. Polymers with narrow molecular-weight distributions, block copolymerizations or sequential polymerizations of THF are possible. Transfer to polymer oxygen occurs and tends to broaden the molecular-weight distribution at longer times and higher temperatures. In additiion, some cyclization reactions also occur (see Chapter 5 and McKenna *et al.*[58]).

4.2.7 Strong protonic acids

The mechanism of THF polymerization with strong protonic acids as initiators is less straightforward than in the case of initiation by superacid esters. The acids that have been studied include HSO_3CF_3, HSO_3F and HSO_3Cl. Once again NMR methods have been used to study these polymerizations, and once again contradictory conclusions have grown out of differences in assignments of the peaks observed in the NMR spectra.[36,59–62] A discussion of kinetics would be premature since NMR assignments and the chemistry of the reactions are still in a state of flux. Because of the industrial interest in protonic-acid initiation of THF polymerization, a brief summary of the present state of knowledge will be included here; more will be said in Chapter 5. In this case the 1978 report of Pruckmayr and Wu[62] seems most likely to be correct, and only these data will be quoted.

Schematically, the important chemical reactions that may occur are shown in Scheme 4.3 with HSO_3F as the initiator. The formation of macroesters from the macroions is not included, but it is the same as for initiation with superacid esters. The principal conclusions of the study are that non-hydrolyzable acids like HSO_3CF_3 lead to PTHF of very high molecular weight as a result of chain-coupling–ring-opening

Initiation

Propagation

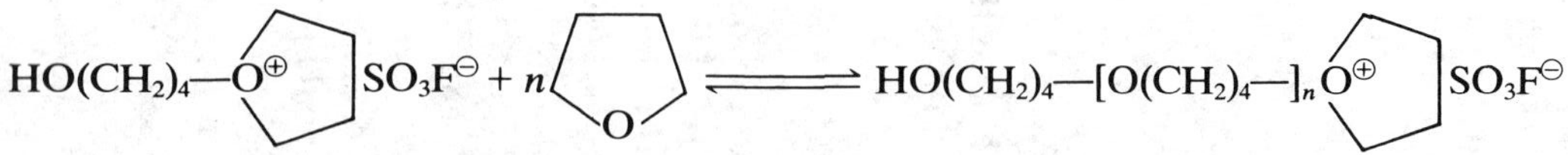

Chain-coupling–Ring-opening Reactions

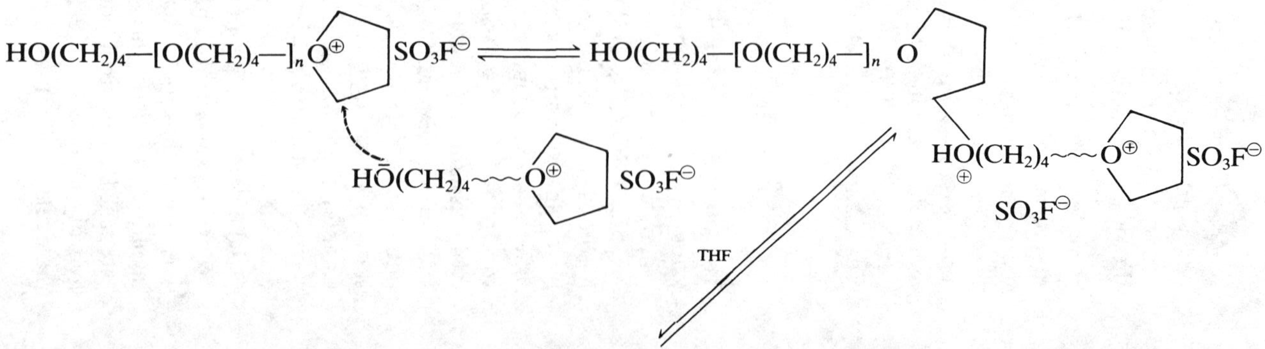

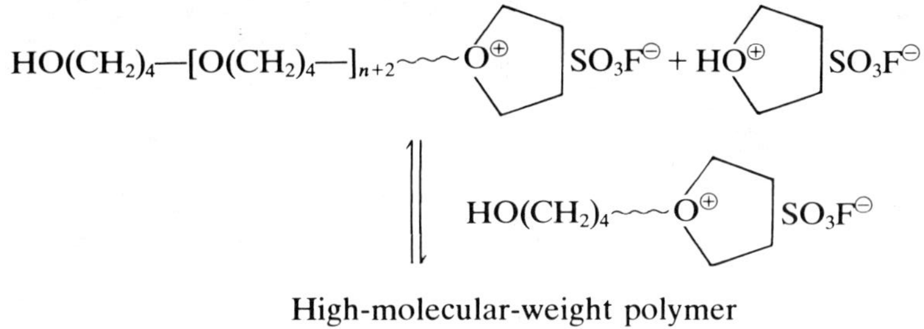

HF-sulfate Formation

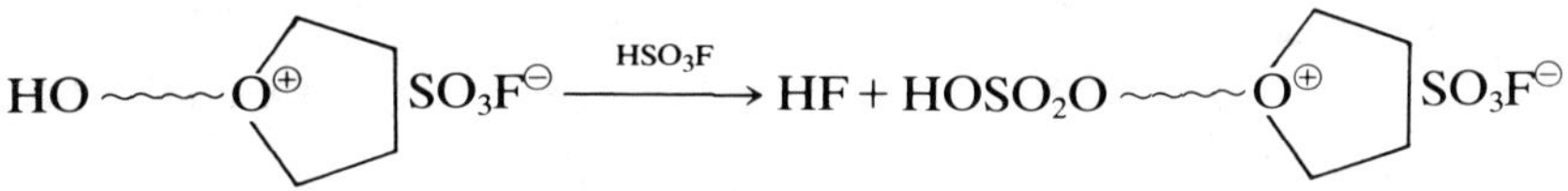

Dialkyl-sulfate Formation

a)

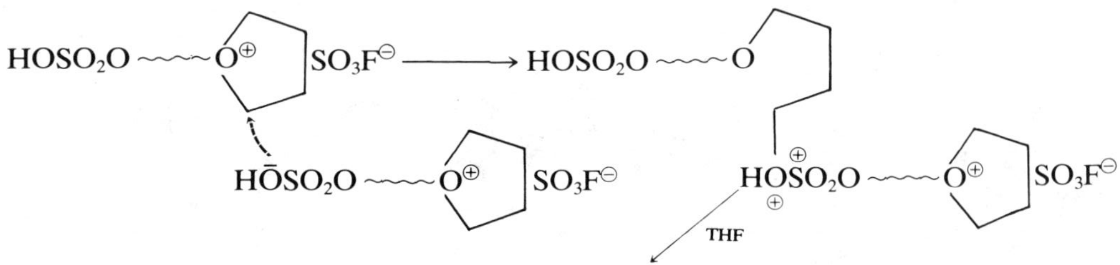

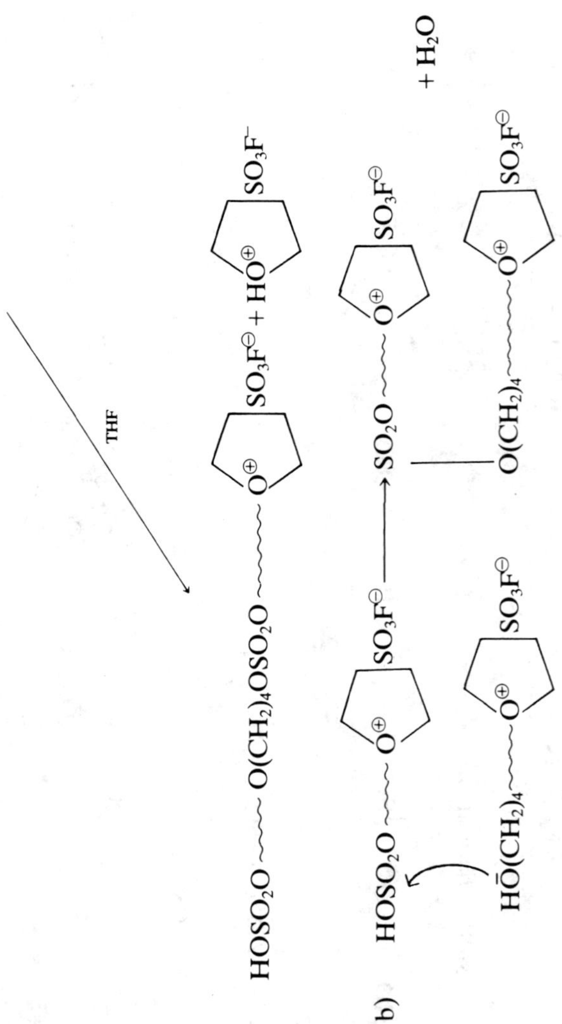

SCHEME 4.3 Chemical reactions in THF polymerizations initiated with HSO_3F.

reactions. Hydrolyzable acids like HSO_3F and HSO_3Cl lead to PTHF of lower molecular weight as a result of the formation of mono- and dialkyl sulfates. HF or HCl is formed as a byproduct. If H_2O is formed as a byproduct by reaction of the hydroxy end group with FSO_3H, H_2SO_4 can also be formed. Sulfate formation is normally irreversible and slower than chain propagation, and the molecular weight goes through a maximum. Alkyl sulfate esters must be hydrolyzed in order to obtain hydroxy-terminated PTHF.[3] Polymerizations with hydrolyzable protonic acids are not "living". Pruckmayr and Wu call them "slowly dying", and state that the rate of termination depends on the polymerization conditions. In the final stages of the polymerization, one or more sulfate groups per molecule will be present. The concentrations of tertiary oxonium ions and of fluorosulfate esters approach zero. Addition of fresh THF at this stage will not lead to continued polymerization.

4.2.8 Electrochemical initiation

Electrochemical initiation of THF polymerization in the presence of salts such as $(C_4H_9)_4N^{\oplus}ClO_4^{\ominus}$ was discussed in Chapter 3, where it was shown that the real initiator is $HClO_4$. In 1971 Nakahama *et al.*[63] reported a kinetic study carried out to elucidate the relationship between the amount of current and the number of propagating ends formed in the electrochemical process. The polymerization was carried out in bulk at 30°C using $(C_4H_9)_4N^{\oplus}ClO_4^{\ominus}$ as an electrolyte. Their experimental results were fitted to the usual THF kinetic equation with [P*] in the form

$$[P^*] = f\left(\frac{Q}{V}\right),$$

where f is the current efficiency; Q is the amount of current; and V is the volume of the system. A reasonable fit to the derived equation was obtained by assuming that one active species was formed by one electron oxidation. The mechanism that they proposed as a result is given in Chapter 3.

4.2.9 Other initiators

Lyudvig *et al.*[25] also reported a study of the kinetics of polymerization of THF in the presence of the reaction product of CH_3COCl and $SbCl_5$, namely, $CH_3CO^+SbCl_6^-$. The rate curves were S-shaped, but there was a point of maximum rate and a steady state was reached. Their result should be compared with the more recent data of Yamashita *et al.*,[21] who showed that initiation with the corresponding SbF_6^- salt is slow (see Table 4.1).

4.3 SUBSTITUTED TETRAHYDROFURANS

Kinetic analyses of the substituted tetrahydrofurans have been made only for bicyclic compounds. For example, Guisti *et al.*[64] and Saegusa *et al.*[65,66] report studies of the polymerization of *exo*- and *endo*-2-methyl-7-oxabicyclo[2.2.1]heptanes.

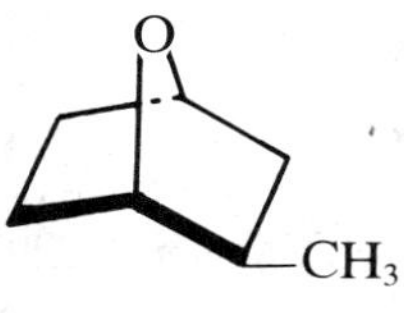

exo

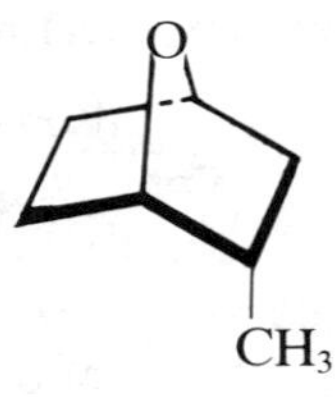

endo

As noted in Chapter 3, it was found that both derivatives lead to polymers in which succeeding ether groups are *trans* to each other. Guisti *et al.* made their kinetic analysis on the basis of the initial velocity observed in the time-conversion curve. As usual, Saegusa *et al.* used their phenoxy end-capping method; the initiator used was the BF_3–ECH system. Polymerizations of both isomers were characterized by slow initiation and fairly rapid termination. In contrast to THF polymerizations, the depolymerization reaction was insignificant and the kinetics were analyzed on the basis of the kinetic equation

$$\ln\{[M]_{t_1}/[M]_{t_2}\} = k\int_{t_1}^{t_2}[P^*]\,dt,$$

in which the $[M]_t$ terms are instantaneous monomer concentrations and $\int_{t_1}^{t_2} [P^*]\,dt$ was obtained by graphical integration. The kinetic parameters are shown in Table 4.8. In this case the difference in kinetic activity can be ascribed to the activation energy, rather than to the frequency factor which was important in the oxetane series. The implication is that the *endo* isomer is more strained than the *exo* isomer and the *endo* isomer thus polymerizes more readily.

Kinetic studies have also been reported for *trans*-2- and *trans*-3-oxabicyclo[3.3.0]octane.[67] Hvilsted and Kops initiated

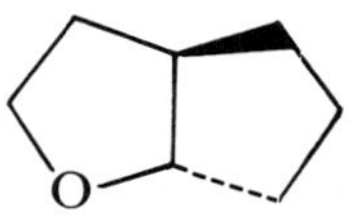

trans-2

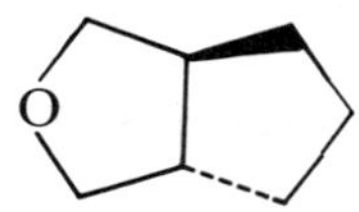

trans-3

the polymerizations in CH_2Cl_2 with triethyloxonium hexafluorophosphate and determined kinetic parameters in the temperature interval from −30 to 0°C.[67] The instantaneous concentration of propagating species was determined by the phenoxy end-capping method. In contrast to the polymerizations of bond-bridged bicyclic ethers, where the heterocyclic ring is fused to a cyclohexane ring, and which are partially or completely reversible, the polymerization of the *trans*-3 five-membered ring compound showed no signs of reversibility. The *trans*-2 compound was somewhat reversible. ^{13}C NMR showed that the repeat units in the polymer correspond to ring opening exclusively by nucleophilic attack of monomer oxygen at the *methylene* carbon adjacent to the oxonium ion of the growing chain end. The mechanism suggested for the *trans*-3 compound was as follows:

$$I + M \xrightarrow{k_i} M^{\oplus}$$

$$M^{\oplus} + M \xrightarrow{k_p} P_n^{\oplus}$$

$$P_n^{\oplus} + M \xrightarrow{k_p} P_{n+1}^{\oplus}$$

TABLE 4.8
Kinetic Parameters for *Exo*- and *Endo*-2-methyl-7-oxabicyclo[2.2.1]heptane[65,66]

Parameter	Temp. °C	*Exo*-MOBH	*Endo*-MOBH
$10^3 \times k_p$, $\mathrm{l\,mol^{-1}\,s^{-1}}$	−40	0.18	0.85
	−30	0.87	4.0
	−20	2.6	8.4
	−10	9.4	25
$\Delta E^{\neq}$, $\mathrm{kJ\,mol^{-1}}$		63	59
$10^{-9} \times A_p^{\neq}$, $\mathrm{l\,mol^{-1}\,s^{-1}}$		41	7

The rate equations were written

$$\ln \{[\mathrm{I}]_{t_1}/[\mathrm{I}]_{t_2}\} = k_i \int_{t_1}^{t_2} [\mathrm{M}^{\oplus}]_t \, dt$$

$$\ln \{[\mathrm{M}]_{t_1}/[\mathrm{M}]_{t_2}\} = k_p \int_{t_1}^{t_2} [\mathrm{P}^{\oplus}] \, dt,$$

where $M^{\oplus}$ and $P^{\oplus}$ refer to the active monomer and polymer species, respectively. Equations for the *trans*-2 compound were the same as those given for THF.

As shown in Table 4.9 the rate of initiation was three orders of magnitude slower than the propagation reaction. This contrasts sharply with the values for THF polymerizations, where the propagation rate is slow compared to the rate of initiation with Et_3OPF_6 (cf. for example, the data for Et_3OBF_4 in Table 4.1; the comparison is reasonable because the data for THF indicate that the rate of reaction of oxonium ions is independent of the counterion). The rate constants for propagation of these bicyclic monomers are large compared to comparable rate constants for THF polymerizations. As shown in Table 4.1, a typical overall rate constant of propagation for THF was $4.6 \times 10^{-3}\ \mathrm{l\,mol^{-1}\,s^{-1}}$. The difference probably results from the greater strain in the bicyclic compounds.

As shown in Table 4.10, activation enthalpies for initiation were higher than those for propagation. As usual, other kinetic

TABLE 4.9
Rate Constants for the Polymerization of[67] *trans*-2- and *trans*-3-Oxabicyclo[3.3.0]octane[a]

Parameter			T, °C	
	0	−10	−20	−30
k_i, 10^3 l mol^{-1} s^{-1}				
trans-2	0.27	0.068	0.019	—
trans-3	—	0.22	0.047	0.014
k_p, l mol^{-1} s^{-1}				
trans-2	0.53	0.14	0.056	—
trans-3	—	0.24	0.075	0.027

[a]In CH_2Cl_2 and initiated by Et_3OPF_6.

TABLE 4.10
Activation Parameters for the Polymerization of[67] *trans*-2- and *trans*-3-Oxabicyclo[3.3.0]octane[a]

Parameter	*trans*-2	*trans*-3
$\Delta E_i^{\neq}$, kJ mol^{-1}	76	73
A_i, l mol^{-1} $^{-1}$	10^{11}	7×10^{10}
$\Delta G_i^{\neq}$, kJ mol^{-1} (−20°C)	85	82
$\Delta H_i^{\neq}$, kJ mol^{-1}	74	71
$\Delta S_i^{\neq}$, J mol^{-1} K^{-1}	−42	−45
$\Delta E_p^{\neq}$, kJ mol^{-1}	64	58
A_p, l mol^{-1} s^{-1}	10^{12}	8×10^{10}
$\Delta G_p^{\neq}$, kJ mol^{-1} (−20°C)	68	67
$\Delta H_p^{\neq}$, kJ mol^{-1}	62	56
$\Delta S_p^{\neq}$, J mol^{-1} K^{-1}	−22	−44

[a]In CH_2Cl_2 and initiated by Et_3OPF_6.

parameters also shown in Table 4.10 were obtained from expressions derived in transition-state theory and were calculated from the relations

$$k_x = (kT/h)\exp\left[-\Delta G_x^{\neq}/RT\right]$$

$$\Delta H_x^{\neq} = \Delta E_x^{\neq} - RT$$

$$\Delta G_x^{\neq} = \Delta H_x^{\neq} - T\Delta S_x^{\neq},$$

where k and h are the Boltzmann and Planck constants, respectively; x refers to either initiation or propagation. The higher values of activation parameters for initiation compared to those for propagation were attributed to relief of strain on ring opening during propagation. No rings are opened during the alkyl exchange occurring on initiation.

GPC traces of the polymers isolated after polymerization of these bicyclic monomers were multimodal. Calculated molecular weights for the lower-molecular-weight fractions suggested the presence of small amounts of trimer and tetramer.

The complete analysis indicated that, although the system appears to be "kinetically living" (i.e., termination reactions are not important), the data cannot be satisfactorily explained without including transfer reactions.

4.4 COPOLYMERS

Copolymerizability of THF with other monomers has already been discussed in Chapter 3. A few points are worth repeating here.

A large number of copolymers of THF have been prepared. Many cyclic compounds that will not homopolymerize do copolymerize readily with THF.[9,68–70] However, the preparation of copolymers does not mean that the chemistry and kinetics of these copolymerizations are well understood. It is known that the copolymerizations proceed by an oxonium-ion mechanism.

All the problems inherent in the interpretation of homopolymerizations are still present, and new problems are generated as a result of the simultaneous presence of two polymerizable monomers and their interaction with each other. The familiar problems from homopolymerizations include doubts about the nature and number of growing centers in a polymerization, solvent effects,[71] catalyst effects, termination reactions that produce low-molecular-weight polymers or cyclic oligomers, relative reactivity of ions formed from monomer and from polymer and transfer reactions of all kinds mentioned before. In addition, some catalysts give polymers with block-copolymer character or even mixtures of homopolymer and copolymer. Backbiting[72] and effects of the penultimate unit have been found.[73]

In studies of the kinetics of copolymerization of THF with other monomers the Mayo–Lewis[74] equations have been applied, often with the appropriate caution but more often without sufficient awareness of the attendant complexities. Many monomer reactivity ratios have been derived in this way; a large number of them have been summarized previously.[9,75] These will not be repeated here, and no attempt to update the lists will be made. Instead, some of the factors that seem to be important in regulating the copolymerizations and some of the newer approaches that have been suggested for dealing with the complicated kinetics will be presented. Only two examples of individual rate studies will be given.

4.4.1 Influence of basicity, ring strain and other factors

Ring strain The factors most commonly mentioned as important in studies of cyclic compounds are relative basicity and ring strain. It seems fairly will established that relative basicity plays an important role in controlling the relative reactivity of cyclic ethers; this lends further support to the belief that propagation proceeds by nucleophilic attack of monomer oxygen in the cationic copolymerizations. In cyclic sulfides, by contrast, basicity seems to be less important than other factors such as ring strain in determining relative reactivity.[76] Some comparisons are shown in Table 4.11.

Basicity Relative basicities of cyclic ethers have been determined by several methods.[68,77–79] In one method, Wirth and Slick[77] obtained relative basicities by infrared spectroscopy from a study of the equilibrium constant at 27°C for the distribution of BF_3 between two ethers in benzene solution. In another case, Yamashita *et al.*,[68] following a method similar to that of Gordy and Stanford,[80] measured the infrared absorption spectra of cyclic ethers in the presence of 0.1 M

TABLE 4.11
Strain Energy and Polymerizability of Cyclic Ethers and Sulfides[a]

Ring size	Strain energy, kJ mole^{-1} Ether	Sulfide	Polymerizability Ether	Sulfide
3	114	83	+	+
4	105	82	+	+
5				
unsubstituted	23	8.4	+	−
substituted	n.a.	n.a.	±	−
bicyclic	n.a.	n.a.	+	−
6	5.0	−1.3	−	−

[a]Most of this data is taken from Stille and Empen[76]; n.a., not available.

methanol-d. They then calculated the basicity of the monomers from the shift value of the O–D stretching band. Nuclear-magnetic-resonance measurements of the equilibrium constant between $BF_3 \cdot Et_2O$ and the cyclic ether were also used.[68]

Yamashita *et al.*[68] point out some features of the basicity of cyclic ethers. The basicity is affected by chemical structure, ring size and substituents. The ring size affects basicity in the order: $4 > 5 > 6 > 3$. In five-membered rings the basicity order is: ether > lactone > formal. Methyl substitution increases the basicity, and chloromethyl substitution decreases basicity. The relative basicities of the monomers are broadly in agreement with their reactivity in copolymerization experiments.[9]

Tanaka[81] has attempted to make a quantitative estimate of the contribution of ring strain and basicity to the reactivity of cyclic ethers in cationic copolymerization. The free energy of polymerization was used as a measure of ring strain. He derived a relationship between the logarithm of relative reactivity, $1/r_n$, of m-membered ring ethers with i substituents to n-membered ring compounds with j substituents and a linear combination of the differences in basicity, $\Delta(pK_b)_{m\cdot i-n\cdot j}$ and in free energy, $\Delta(\Delta G)_{m\cdot i-n\cdot j}$.

$$\log(1/r_n) = \alpha\Delta(\Delta G)_{m\cdot i-n\cdot j} + \beta\Delta(pK_b)_{m\cdot i-n\cdot j} + \gamma\,,$$

where α, β and γ are constants. He compared calculated values of $1/r_n$ with observed values taken from the literature, and found good agreement.

Ring size Saegusa *et al.*[28,82] have compared the kinetic parameters determined for cyclic oxides of different ring sizes. Their comparisons are shown in Table 4.12. All of these parameters were determined under similar conditions and using the same techniques, so that the comparison is significant. The order of reactivity is: oxetane > THF > oxepane. Saegusa *et al.* attribute the difference in reactivity of oxetane and THF to a difference in frequency factor. Both rings are essentially planar and strained, but the oxetane ring is more strained. In contrast, the difference in the reactivity of THF and oxepane seems to come from the larger activation energy for oxepane. The reason for the larger activation energy may again be ring strain. The strain in a seven-membered-ring oxonium ion can be relieved by small deformations of the

TABLE 4.12
Propagation Rate Coefficients and Activation Parameters of[28,82] Cyclic Ether Polymerizations[a]

Parameter	Temp., °C	Ring size 4	5	7
$k_p \times 10^3$, $l\,mol^{-1}\,s^{-1}$	−28	7.5	—	—
	−23	13	—	—
	−10	57	1.7	0.0033
	0	140	4.1	0.015
	10	—	8.4	0.031
	20	—	—	0.091
	30	—	—	0.43
$\Delta G^{\neq}$, $kJ\,mol^{-1}$ at 0°C		71	92	113
$\Delta E_p^{\neq}$, $kJ\,mol^{-1}$		59	50	75
$10^{-7} A_p^{\neq}$, $l\,mol^{-1}\,s^{-1}$		5,300	1.1	190

[a]Solution polymerization in CH_2Cl_2. For oxetane, $[M]_0 = 3.1$ M, $[BF_3{\cdot}THF]_0 = 0.003$ M; for THF, $[M]_0 = 6.3$ M, $[BF_3{\cdot}THF]_0 = [ECH]_0 = 0.01$ M; for oxepane, $[M]_0 = 2.9$ M, $[BF_3{\cdot}THF]_0 = [ECH]_0 = 0.057$ M.

angles to other bonds. It is interesting to note that the free energy increases in the order: $4<5<7$. From these data it seems that activation energy and frequency factor are better measures of reactivity than free energy.

4.4.2 Individual Rate Studies

Propylene oxide, epichlorohydrin, glycidic acid nitrate Korovina *et al.*[83] have investigated the polymerization and copolymerization of THF, propylene oxide, epichlorohydrin and glycidic acid nitrate using BF_3 and BF_3 complexes as initiator. CH_2Cl_2, CCl_4, $C_6H_5NO_2$ and bulk THF were used as solvents. Small additions of polyoxypropylene glycol, $NaClO_4$ or water were made to some of the polymerizations. They conclude that a change in the conditions of solvation of reaction sites of the components of the system leads to significant changes in the course of the polymerization. The yields of cyclic tetramer arising from the presence of epoxide and the molecular weights of the resulting copolymers could be changed by changing the medium. The rates of copolymerization and the sequence distribution of the monomers in the copolymers could be altered by the additives. The results are interpreted in terms of an active site formed by oppositely charged ends of a Zwitterion with an $-OBF_3^-$ end group rather than the BF_4^- counterion shown in Chapter 3 for polymerizations initiated by BF_3.

3,3-bis-Chloromethyloxetane (*BCMO*). Copolymerization of THF with BCMO is probably the best-understood system from a kinetics viewpoint. Two rather careful studies have been made, one above and the other below the ceiling temperature of polymerization for THF. Both will be described here, because they illustrate the complexities of even a seemingly simple copolymerization that can be assumed to be essentially free from transfer and termination reactions.

Copolymerization below T_c Yamashita *et al.* have derived a copolymer composition equation that includes a depropagation reaction which might be expected in the cationic copolymerization of BCMO and THF.[84] They consider two models. For the first one it is assumed that monomer M_2 adds reversibly to both active chain ends m_1^* and m_2^* and that depropagation by detachment of an M_1 unit can be neglected the elementary reactions are then

$$(1)\quad \sim m_1^* + M_1 \xrightarrow{k_{11}} \sim m_1 m_1^*$$

$$(2)\quad \sim m_1^* + M_2 \underset{k'_{12}}{\overset{k_{12}}{\rightleftharpoons}} \sim m_1 m_2^*$$

$$(3)\quad \sim m_2^* + M_1 \xrightarrow{k_{21}} \sim m_2 m_1^*$$

$$(4)\quad \sim m_2^* + M_2 \underset{k'_{22}}{\overset{k_{22}}{\rightleftharpoons}} \sim m_2 m_2^*$$

They define the parameters r_1, r_2, ϕ, ρ, δ and α as follows:

$$r_1 = k_{11}/k_{12}\,,$$

$$r_2 = k_{22}/k_{21}\,,$$

$$\phi = k_{22}/k'_{22}\,,$$

$$\rho = k_{12}/k'_{12}$$

$$\delta = k_{11}/k_{22}\,,$$

$$\alpha = [(m_2)^*_{n+1}]/[(m_2)^*_n]\ (n \geq 1)\,.$$

The final copolymer composition is given by

$$\left\{\frac{\{M_1\}}{\{M_1\}+\{M_2\}}\right\}_{\text{copoly}} = \frac{(1-\alpha)+r_1(1-\alpha)[M_1]/[M_2]+(1-\alpha)^2\delta r_2/\rho[M_2]}{(2-\alpha)+r_1(1-\alpha)[M_1]/[M_2]+(1-\alpha)^2\delta r_2/\rho[M_2]}\,.$$

The second model was similar, except that both M_1 and M_2 monomer add to the active chain end m_2^* reversibly, and depropagation of the active end m_1^* is neglected. That is, elementary reactions (1) and (2) are considered to be irreversible, and elementary reactions (3) and (4) are considered to be reversible. Two additional parameters are defined:

$$\sigma = k'_{21}/k_{21}\,,$$

$$\beta = \frac{[(m_2)_n(m_1)_1^*]}{[(m_2)_n^*]}\ (n \geq 1)\,,$$

where $[(m_2)_n(m_1)_1^*]$ denotes the concentration of chain ends which have a terminal M_1 unit with n units of M_2 preceding it, i.e.

$$\sim m_1\underbrace{m_2 - - - m_2}_{n}m_1\sim$$

The final equation is then

$$\left\{\frac{\{M_1\}}{\{M_1\}+\{M_2\}}\right\}_{\text{copoly}} = \frac{(1-\alpha)(1+r_1X)}{(1-\alpha)(1+r_1X)+1}$$

where $X = [M_1]/[M_2]$.

By assigning M_1 as BCMO and M_2 as THF, Yamashita *et al.* obtain a good fit to the second model, for polymerizations conducted at 30–50°C with $Et_3O^+BF_4^-$ as catalyst and in CH_2Cl_2 solution. They conclude that their study demonstrates the effect of the penultimate THF monomer unit on the depropagation reaction. It should be noted that Yamashita *et al.* do not attempt to determine the concentration of growing chains or the degree of termination occurring in their copolymerizations. With the BF_4^- counterion, substantial termination would be expected for polymerization at 50°C. They do report a definite effect of initiator and counterion on the course of the copolymerization, and point out that different initiators lead to varying degrees of alternating copolymerization.

Copolymerization above T_c This study includes not only BCMO but also other substituted oxetanes. Penczek and Kubisa[85] studied the copolymerization of oxetanes with a large excess of THF above the ceiling temperature of THF. The catalyst used was triisobutylaluminum, and only monomers were used as solvent. Highly ordered copolymers were formed. If the oxetane was 3,3-dimethyloxetane (DMO) or 3,3-bis(fluoromethyl)oxetane (BFMO), the limiting composition [thf]/[ox] of the polymer tended toward one mole of THF for every mole of oxetane as the excess of THF monomer grew larger. If the oxetane was BCMO, the limiting composition was close to two moles of THF for every mole of oxetane. In order to explain their results, Kubisa and Penczek recall the reversibility of ionic copolymerizations, especially above the ceiling temperature of one of the monomers. They suggest that, if a depropagation reaction involving a THF molecule is to occur, a transition state such as is shown in the diagram needs to form:

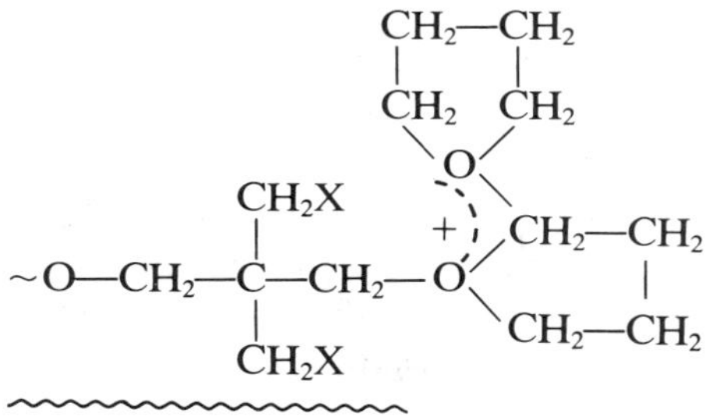

where X = H, F or Cl. Thus, before a monomer molecule splits off the chain end, a partial bond within a penultimate unit is formed in the transition state. If the antepenultimate unit in the depropagation step, underlined by a wavy line in the formula, does not disturb the properties of the active centers which have THF molecules at their end, there is no reason to expect any interference with the propagation step. This seems to be the case when X = H or F. However, in the copolymerization of the BCMO–THF pair, the influence of the penultimate unit structure (or antepenultimate unit in the depropagation step) cannot be ignored. Analysis of Stuart molecular models showed that the chloromethyl groups of BCMO in these units are frozen and cannot rotate freely. Thus the depropagation reaction proceeds much more slowly in this case. Thermodynamically, the depropagation reaction proceeds with a much more negative entropy of activation than the same reaction with the other monomers. The kinetics of the copolymerization are also affected. The following scheme was proposed for these polymerizations:

$$(1)\quad \sim m_1^* + M_1 \xrightarrow{k_{11}} \sim m_1^*$$

$$(2)\quad \sim m_1^* + M_2 \xrightarrow{k_{12}} \sim m_1 m_2^*$$

$$(3)\quad \sim m_1 m_2^* + M_2 \underset{k_{-22}}{\overset{k_{22}}{\rightleftharpoons}} \sim m_2 m_2^*$$

$$(4)\quad \sim m_1 m_2 m_2^* + M_2 \underset{k_{-222}}{\overset{k_{222}}{\rightleftharpoons}} \sim m_2 m_2 m_2^*$$

$$(5)\quad \sim m_1(m_2)_n m_2^* + M_1 \longrightarrow \sim m_1(m_2)_{n+1} m_1^*, \quad n \geq 0\,.$$

The symbols m_2 and M_2 always refer to the THF polymer and monomer, respectively. Positive values of k refer to the forward reactions and negative values of k to reverse reactions, as usual. The scheme is based on several facts and assumptions, as follows.

i) The chain propagation of oxetanes is considered to be practically irreversible. This is a reasonable assumption based on experience with oxetanes in homopolymerizations.

ii) Addition of a THF molecule to an active center bearing an oxetane at its end is irreversible, since depropagation would require cyclization to a four-membered ring.

iii) Addition of THF to its own active center is reversible.

iv) Addition of oxetane to a THF active center is irreversible.

Some elements of the scheme were previously presented by

Yamashita *et al.*[84] The general solution of the scheme is given by the equation according to Kubisa and Penczek.[85]

$$\frac{m_1}{m_2} = \frac{\{(r_1[M_1])/[M_2] + 1\}(1 + \alpha + \beta)}{1 + 2\alpha + 3\beta},$$

where

$$\alpha = \frac{[m_1m_2m_2^*]}{[m_1m_2^*]} = \frac{[M_2]}{1/K_1 + (1/r_2)[M_1] + K_2^2 r_2[M_2]/(r_2 + K_2[M_1])}$$

$$\beta = \frac{[\sim m_1m_2m_2m_2^*]}{[-m_1m_2m_2^*]} = \frac{[M_2]}{1/K_2 + (1/r_2)[M_1]}$$

$$r_1 = k_{11}/k_{12}$$

$$r_2 = k_{22}/k_{21}\ ;$$

K_1 is the equilibrium constant of the propagation–depropagation equilibrium of a sequence such as –(bcmo)–(thf)– 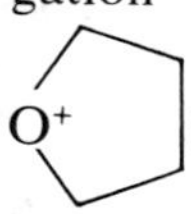; and K_2 is the equilibrium constant of the propagation–depropagation equilibrium of a sequence such as –(thf)–(thf)– 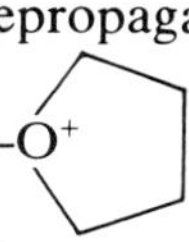. For the copolymerization involving DMO and BFMO, $\alpha = \beta = 0$ and $K_1 = K_2 = 0$. Hence the general solution reduces to

$$\frac{m_1}{m_2} = 1 + r_1\frac{[M_1]}{[M_2]}.$$

Thus r_1 can be determined from a plot of m_1/m_2 against $[M_1]/[M_2]$. For the copolymerization involving BCMO, $\beta = 0$ and $K_2 = 0$, but $\alpha \neq 0$. In this case r_1 was measured as the slope of the tangent to the experimental plot relating m_1/m_2 and $[M_1]/[M_2]$ when $[M_2]/[M_1] \to 0$. To determine K_1 and r_2, a knowledge of the dependence of α on $[M_1]/[M_2]$ is required. If α is written as a function of $[M_1]/[M_2]$ and m_1/m_2, that is,

$$\alpha = \frac{(m_1/m_2) - 1 - r_1[M_1]/[M_2]}{2(m_1/m_2) - 1 - r_1[M_1]/[M_2]},$$

the values of α can be determined from the same plot that gave r_1. Furthermore,

$$\frac{[M_2]}{\alpha} = \frac{1}{K_1} + \frac{[M_1]}{r_2}.$$

Thus a plot of $[M_2]/\alpha$ against $[M_1]$ allows the value of K_1 to be determined from the intercept and r_2 from the slope. The resulting kinetic parameters are summarized in Table 4.13. The comparison with the basicity values, here given in terms of pK_b, again shows that, in a copolymerization of cyclic ethers, reactivity correlates well with basicity.

TABLE 4.13
Kinetic Parameters in Copolymerization of Oxetanes and THF[a] and Basicities of Monomer[85]

Oxetane	r_1	r_2	K_1, M	K_2, M	pK_b
DMO	4.5	—	0	0	3.13[b]
BFMO	0.25	—	0	0	—
BCMO	0.35	9.0	0.8	0	5.65

[a]For copolymerizations carried out above T_c for THF.

[b]Taken to be approximately equal to the value for oxetane. Note that the pK_b for THF is 5.00.

4.4.3 Concluding comments

Although copolymerizations of THF with many comonomers have been reported, really meaningful determination of the kinetic parameters is still in its infancy. Saegusa *et al.*[86] have even suggested that the apparent values of the monomer reactivity ratios in cyclic-ether copolymerizations may be more influenced by the exchange given below than by kinetic reactivity of the ring-opening process of a propagation reaction

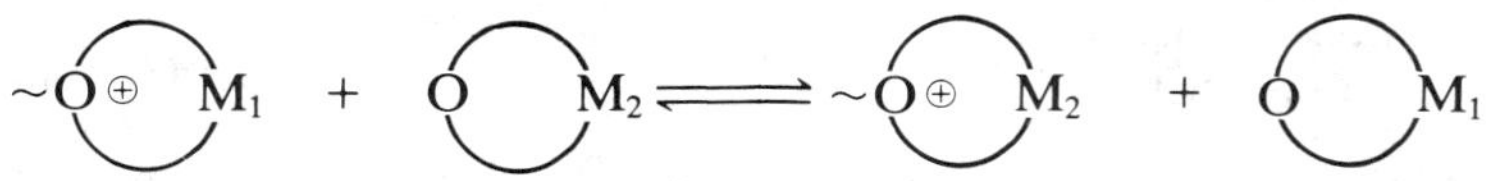

Also, the data of Kubisa and Penczek[85] in the copolymerization of THF and substituted oxetanes show that the depropagation reaction may play an important role in copolymerization kinetics. Even penultimate and antepenultimate units may be important.

REFERENCES

1. M. Szwarc, *Macromolecules*, **11**, 1053 (1978).
2. S. Penczek, *Macromolecules*, **12**, 1010 (1979).
3. H. Meerwein, D. Delfs and H. Morshel, *Angew. Chem.*, **72**, 927 (1960).
4. J. Furukawa and T. Saegusa, "Polymerization of Aldehydes and Oxides", Interscience, New York, NY (1963).
5. P. Dreyfuss and M. P. Dreyfuss, *Adv. Polym. Sci.*, **4**, 528 (1967).
6. T. Saegusa and S. Kobayashi, *Prog. Polym. Sci. Jap.*, **6**, 107 (1973).
7. N. S. Enikolopian, V. V. Ivanov, G. V. Koronina, M. A. Markevich, T. I. Prokof'eva, A. T. Ponomarenko and G. V. Rakova, *Vysokomol. Soyed.*, **A19**, 1924 (1977) (in Russian). Translated in *Polym. Sci. USSR*, **19**, 2200 (1977).
8. A. Ledwith and C. Fitzsimmonds, in "Polymer Chemistry of Synthetic Elastomers", Part I, J. P. Kennedy and E. G. M. Törnqvist (eds.), Interscience, New York, NY, Chap. 5C, pp. 377–418 (1968).
9. P. Dreyfuss and M. P. Dreyfuss, in "Ring-opening Polymerization", K. C. Frisch and S. L. Reegen (eds.), Marcel Dekker, New York, NY, Chap. 2 (1969).
10. P. Dreyfuss and M. P. Dreyfuss, in "Comprehensive Chemical Kinetics", Vol. 15, C. H. Bamford and C. F. H. Tipper (eds.), Elsevier, Amsterdam-W, Netherlands, Chap. 4 (1976).
11. T. Saegusa and S. Kobayashi, *ACS Symposium Series*, "Polyethers", **6**, 150 (1975).
12. G. Pruckmayr and T. K. Wu, *ACS Symposium Series*, "Carbon-13 NMR in Polymer Science", **103**, 237 (1979).
13. S. Penczek, *Makromol. Chem. Suppl.*, **3**, 17 (1979).
14. T. Saegusa and S. Matsumoto, *J. Polym. Sci.: A1*, **6**, 1559 (1968).
15. P. Dreyfuss and M. P. Dreyfuss, *Adv. Chem. Series*, **91**, 335 (1969).

16. D. Vofsi and A. V. Tobolsky, *J. Polym. Sci.*, **A3**, 3261 (1965).
17. T. Saegusa and S. Matsumoto, *J. Macromol. Sci.-Chem.*, **A4**, 873 (1970).
18. S. Penczek, in "Advances in Ionic Polymerization, Proceedings of International Symposium", p. 147 (1972); *CA*, **84**, 180665 (1976).
19. S. Słomkowski and S. Penczek, *J. Chem. Soc., Perkin Trans.*, **2**(14), 1718 (1974).
20. G. Heublein, S. Spange and P. Hallpup, *Makromol. Chem.*, **180**, 1935 (1979).
21. Y. Yamashita, S. Kozawa, M. Hirota, K. Chiba, A. Matsui, H. Hirao, M. Kadama and K. Ito, *Makromol. Chem.*, **142**, 171 (1971).
22. Y. Yamashita, H. Matsui, G. Hattori, S. Kozawa and M. Hirota, *Makromol. Chem.*, **142**, 183 (1971).
23. B. A. Rozenberg, O. M. Chekhuta, E. B. Lyudvig, A. R. Gantmakher and S. S. Medvedev, *Polym. Sci. USSR*, **6**, 2246 (1964).
24. J. M. Sangster and D. J. Worsfold, *Macromolecules*, **5**, 229 (1972).
25. E. B. Lyudvig, B. A. Rozenberg, T. M. Zuereva, A. R. Gantmakher and S. S. Medvedev, *Polym. Sci. USSR*, **7**, 296 (1965).
26. J. Stejny, *J. Macromol. Sci.-Chem.*, **A7**, 1435 (1973).
27. T. Saegusa and S. Matsumoto, *Macromolecules*, **1**, 442 (1968).
28. T. Saegusa, T. Shiota, S. Matsumoto and H. Fujii, *Macromolecules*, **5**, 34 (1972).
29. R. A. Barzykina, G. N. Komratov, G. V. Korovina and S. G. Entelis, *Polym. Sci. USSR*, **20**, 197 (1978).
30. R. A. Barzykina, G. N. Kamratov, N. G. Taganov, G. V. Korovina and S. G. Entelis, *Polym. Sci. USSR*, **19**, 844 (1977).
31. H. Imai, T. Saegusa, S. Matsumoto, T. Tadasa and J. Furukawa, *Makromol. Chem.*, **102**, 222 (1967).
32. T. Saegusa, H. Imai and S. Matsumoto, *J. Polym. Sci.: A1*, **6**, 459 (1968).
33. T. Saegusa, S. Matsumoto and Y. Hoshimoto, *Polym. J.*, **1**, 31 (1970).
34. S. Smith and A. Hubin, *J. Macromol. Sci.-Chem.*, **A7**, 1399 (1973).
35. S. Kobayashi, H. Danda and T. Saegusa, *Macromolecules*, **7**, 415 (1974).
36. G. Pruckmayr and T. K. Wu, *Macromolecules*, **6**, 33 (1973).
37. T. K. Wu and G. Pruckmayr, *Macromolecules*, **8**, 77 (1975).
38. G. Pruckmayr and T. K. Wu, *Macromolecules*, **8**, 954 (1975).
39. K. Matyjaszewski, P. Kubisa, and S. Penczek, *J. Polym. Sci.: Polym. Chem. Ed.*, **12**, 1333 (1974).
40. K. Matyjaszewski and S. Penczek, *J. Polym. Sci.: Polym. Chem. Ed.*, **12**, 1905 (1974).
41. K. Matyjaszewski, P. Kubisa and S. Penczek, *J. Polym. Sci.: Polym. Chem. Ed.*, **13**, 763 (1975).
42. K. Matyjaszewski, A. M. Buyle and S. Penczek, *J. Polym. Sci., Polym. Lett. Ed.*, **14**, 125 (1976).
43. A. M. Buyle, K. Matyjaszewski and S. Penczek, *Macromolecules*, **10**, 269 (1977).
44. K. Matyjaszewski and S. Penczek, *J. Polym. Sci.: Polym. Chem. Ed.*, **15**, 247 (1977).

45. K. Matyjaszewski, S. Słomkowski and S. Penczek, *J. Polym. Sci.: Polym. Chem. Ed.*, **17**, 69 (1979).
46. K. Matyjaszewski, S. Słomkowski and S. Penczek, *J. Polym. Sci.: Polym. Chem. Ed.*, **17**, 2413 (1979).
47. K. Matyjaszewski, T. Diem and S. Penczek, *Makromol. Chem.*, **180**, 1817 (1979).
48. K. Matyjaszewski, S. Penczek and E. Franta, *Polymer*, **20**, 1184 (1979).
49. John A. Dean (ed.), *Lange's Handbook of Chemistry*, 11th edn, McGraw-Hill, New York, NY, p. 10-150 (1973).
50. J. Leonard and D. Maheux, *J. Macromol. Sci.-Chem.*, **A7**, 1421 (1973).
51. S. Penczek and K. Matyjaszewski, *J. Polym. Sci.: Polym. Symp.*, **56**, 255 (1976).
52. P. Dreyfuss and J. P. Kennedy, *J. Appl. Polym. Sci.: Appl. Polym. Symp.*, **30**, 165 (1977).
53. S. Smith, W. J. Schultz and R. A. Newmark, *Polym. Preprints*, **18**(1), 6 (1977); *ACS Symposium Series*, "Ring-opening Polymerization", **59**, 13 (1977).
54. M. Szwarc, "Ions and Ion Pairs in Organic Reactions", Wiley, New York, NY, p. 410 (1968).
55. M. Szwarc and J. J. Hermans, *J. Polym. Sci. B*, **2**, 815 (1964).
56. P. Bourdanducq and D. J. Worsfold, *Macromolecules*, **8**, 562 (1975).
57. K. Matyjaszewski and M. Zielinski, *J. Macromol. Sci.-Chem.*, **A13**, 193 (1979).
58. J. M. McKenna, T. K. Wu and G. Pruckmayr, *Macromolecules*, **10**, 877 (1977).
59. E. B. Lyudvig, E. L. Berman, Z. N. Nysenko, V. A. Ponomarenko and S. S. Medvedev, *Vysokomol. Soyed.*, **A13**, 1375 (1971) (in Russian). Translated in *Polym. Sci. USSR*, **13**, 1546 (1971).
60. K. Matyjaszewski and S. Penczek, *Macromolecules*, **7**, 136 (1974).
61. M. Ú. Mahmud, G. Wegner, W. Kern, J. B. Lando and Y. Osada, *J. Macromol. Sci.-Chem.*, **A11**, 2233 (1977).
62. G. Pruckmayr and T. K. Wu, *Macromolecules*, **11**, 662 (1978).
63. S. Nakahama, S. Hino and N. Yamazaki, *Polymer J.*, **2**, 56 (1971).
64. M. Baccaredda, P. Guisti, F. Andruzzi, P. Cerrai and M. Dimaina, *J. Polym. Sci.*, **C31**, 157 (1970).
65. T. Saegusa, S. Matsumoto, M. Motoi and H. Fujii, *Macromolecules*, **5**, 236 (1972).
66. T. Saegusa, S. Matsumoto, M. Motoi and H. Fujii, *Macromolecules*, **5**, 815 (1972).
67. S. Hvilsted and J. Kops, *Macromolecules*, **12**, 889 (1979).
68. Y. Yamashita, T. Tsuda, M. Okada and S. Iwatsuki, *J. Polym. Sci.: A1*, **4**, 2121 (1966).
69. T. Tsuda, T. Nomura and Y. Yamashita, *Makromol. Chem.*, **86**, 301 (1965).
70. M. P. Dreyfuss and P. Dreyfuss, *J. Polym. Sci.: A1*, **4**, 2179 (1966).
71. E. J. Alvarez, V. Hornof and L. P. Blanchard, *J. Polym. Sci.: A1*, **10**, 2237 (1972).

72. Y. Yamashita, F. Inoue, G. Hattori and K. Ito, *Makromol. Chem.*, **151**, 91 (1972).
73. J. Furukawa and K. Tada, in "Ring-opening Polymerization", K. C. Frisch and S. L. Reegen (eds.), Marcel Dekker, New York, NY, Chap. 3 (1969).
74. F. R. Mayo and F. F. Lewis, *J. Am. Chem. Soc.*, **66**, 1594 (1944).
75. P. Dreyfuss and M. P. Dreyfuss, *Encyclopedia Polym. Sci. Tech.*, **13**, 670 (1970).
76. J. K. Stille and J. A. Empen, *J. Polym. Sci.: A1*, **5**, 273 (1967).
77. H. E. Wirth and P. L. Slick, *J. Phys. Chem.*, **66**, 2277 (1962).
78. S. Iwatsuki, N. Takigawa, M. Okada, Y. Yamashita and Y. Ishii, *J. Polym. Sci. B*, **2**, 549 (1964).
79. R. Pollet, *Ind. Chim. Belge*, **32**(1), 32 (1967).
80. W. Gordy and S. C. Stanford, *J. Chem. Phys.*, **7**, 93 (1939); **8**, 170 (1940); **9**, 204, 215 (1941).
81. Y. Tanaka, *J. Macromol. Sci.-Chem.*, **A1**, 1059 (1967).
82. T. Saegusa, *J. Macromol. Sci.-Chem.*, **A6**, 997 (1972).
83. G. V. Korovina, D. D. Novikov, D. Ya. Rossina, T. V. Grinevich, N. G. Taganov, R. A. Barzykina, G. N. Kamratov and S. G. Entelis, *Vysokomol. Soyed.*, **A18**, 1253 (1976) (in Russian). Translated in *Polym. Sci. USSR*, **18**, 1438 (1976).
84. Y. Yamashita, H. Kasahara, K. Suyama and M. Okada, *Makromol. Chem.*, **117**, 242 (1968).
85. P. Kubisa and S. Penczek, *J. Macromol. Sci.-Chem.*, **A7**, 1509 (1973).
86. T. Saegusa, H. Fujii, S. Kobayashi, H. Ando and R. Kawase, *Macromolecules*, **6**, 26 (1973).

CHAPTER 5

Solution Properties

5.1 INTRODUCTION

PTHF is a saturated, linear, partially crystalline polymer. Under most polymerization conditions, the formation of even occasional branches is an usual occurrence. In addition, PTHF has a repeat unit with four methylene groups and one oxygen atom, $[(CH_2)_4O–]$. Its structure causes it to be soluble in a wide variety of solvents. Discussion of the properties of PTHF in solution is the subject of the present chapter. Other properties of PTHF are described in Chapter 6. The properties of the industrially important polyurethanes and polyesters containing PTHF and of the glycol precursor are presented in Chapter 7.

5.2 SOLVENTS

The presence of the ether oxygen as every fifth atom along the backbone of PTHF greatly increases the solubility of PTHF in polar solvents compared to hydrocarbon polymers. Solubility varies with molecular weight but there are solvents that will dissolve PTHF of all molecular weights. As usual, lower-molecular-weight polymers dissolve more readily in more solvents and form less viscous solutions. Solvents for low-molecular-weight (600–3000) PTHF glycols are listed in Chapter 7. The solubility of other low-molecular-weight PTHFs can be assumed to be the same. Typical solvents for nearly all molecular weights include benzene, toluene, chlorobenzene, nitrobenzene, CH_2Cl_2, $CHCl_3$, CCl_4, CH_3NO_2, THF, ethyl acetate and other esters and liquid SO_2. Ether, alcohols, acetone and even water are solvents for lower-molecular-weight PTHF, although PTHF of even low molecular weight will precipitate in an ice-water slurry, because under those conditions the PTHF crystallizes.

Nonsolvents are aliphatic hydrocarbons such as pentane, petroleum ether and hexane. Selection of a poor solvent for a polymer is often desirable when studying solution properties, because higher concentrations can be used and effects of

TABLE 5.1
Theta Solvents for PTHF

Solvent	Composition[a]	Theta temperature,[b] °C	Method[c]	Ref.
Toluene	—	−28 ± 3	d	1
n-Butanol	—	0–10	e	2
Chlorobenzene/*n*-octane	25.0/75.0	10.0	f	3
Chlorobenzene	—	25	g	1
Benzene/acetonitrile	38.5/61.5	25	g	1
Butanone/acetonitrile	61.7/38.3	25	f	1
Carbon tetrachloride/acetonitrile	49.8/50.2	25	f	1
Chlorobenzene/acetonitrile	39.9/60.1	25	f	1
Tetrahydrofuran/acetonitrile	41.3/58.7	25	f	1
Toluene/acetonitrile	39/61	25	f	1
Chlorobenzene/*n*-octane	21.5/78.5	25.8	f	3
Heptane/cyclohexane	39.2/61.8	26	h	4
Ethyl acetate/*n*-hexane	22.7/77.3[i]	31.8	e	5
Ethyl acetate/*n*-hexane	22.3/77.7[i]	30.4	j	6
Diethyl malonate	—	33.5	e	6
Isopropanol	—	44.6	e	6
Chlorobenzene/*n*-octane	14.5/85.5	46.0	f	3
Chlorobenzene/*n*-octane	13.0/87.0	63.0	f	3
Chlorobenzene/*n*-octane	10.9/89.1	80.0	f	3

[a]On a volume/volume basis except as otherwise noted.
[b]Theta temperatures of endothermic solutions.
[c]All methods are described in Ref. 1, pp. (IV-158)–(IV-159).
[d]Second virial coefficient from osmotic pressure.
[e]Phase equilibria.
[f]Cloud-point titration.
[g]Viscosity–molecular-weight relationship.
[h]Visual observations.
[i]Expressed in wt.%.
[j]Second virial coefficient from light scattering.

nonideality are minimized. Theta solvents are most suitable for theoretical work. Table 5.1 lists a variety of theta solvents for PTHF.

The following list of solvents covers the range from thermodynamically poor to thermodynamically good and is indicative of the type of solvent that falls into each category: diethyl malonate, methyl ethyl ketone, *n*-propyl acetate, *n*-butanol, THF, chlorobenzene and toluene.[7]

5.3 FRACTIONATION

PTHF having a narrow molecular-weight distribution can be prepared.[8] Recently, there have been claims that high yields of polymer with any desired molecular weight are accessible.[9] However, unless special procedures are used, most synthetic methods that give polymers of moderate to high molecular weight yield polymers with a normal distribution. Therefore, fractionation is sometimes carried out when a range of molecular weights of PTHF with a narrow distribution is desired.

Makletsova *et al.*[10] separated PTHF into fractions by successive precipitation from toluene with methanol. Rozenberg *et al.*[11] obtained fractional precipitation by lowering the temperature of solutions of PTHF in acetone. Ali and Huglin[12] collected fractions after incremental addition of MeOH to a 1.5% weight/volume solution of PTHF in benzene at 24°C. A few years later Evans and Huglin[6] carried out a similar fractionation at 25°C on a 0.5% solution of PTHF in benzene, but they added 0.1% 2,6-di-*tert*-butyl-*p*-cresol stabilizer (antioxidant) to the benzene, and stored the fractions in benzene solutions containing 0.2% antioxidant. The antioxidant was added to prevent the steady degradation in air of PTHF in benzene that had previously been reported by Davis and Golden.[13] (Degradation of PTHF is discussed in Chapter 6.) Fractionation of low-molecular-weight PTHF glycols (1,200–9,400) at 22°C using diethyl ether as solvent and petroleum ether as nonsolvent has also been reported.[14] Toluene as solvent and methanol as nonsolvent at 20°C has been used for fractionating PTHF with a molecular weight of about 6×10^4.[7]

5.4 INTRINSIC VISCOSITY

5.4.1 Effect of solvent

Mark–Houwink equations Intrinsic viscosity is a property very often cited in the characterization of PTHF. In early work, the majority of authors quoted values determined in benzene or toluene, usually at 25°C. However, more recently, and especially for theoretical work where theta solvents are preferred, other solvents have been used, and constants for the Mark–Houwink equation, $[\eta] = KM^a$, for a wide variety of solvents have been determined. In the equation, $[\eta]$ is the intrinsic viscosity, M is the molecular weight and K and a are constants. Many of these constants are tabulated in Table 5.2. Constants for theta solvents were $a = 0.5$ are included. The equations should be used with the usual precautions. Those for nontheta solvents are probably valid only in the molecular-weight range given, and probably only for molecular-weight distributions similar to those used by the investigators cited. The data for benzene suggest that the values of the constants are also quite temperature-sensitive. There is disagreement in the literature about whether or not the K value for theta solvents has a unique value. Elias and Adank[20] report a value of 0.202 ml/g for several solvents. Ali and Huglin[12] report changes with solvent, and suggest that the differences may be explained on the basis of a thermal variation of short-range interactions. With these reservations, the equations for theta solvents can probably be used over the whole molecular-weight range as long as temperature is carefully controlled.

Solvent mixtures near the critical point Another interesting result of the effect of solvent on the viscosity of PTHF was obtained by Dondos and Izumi.[21] They found that the intrinsic viscosity of PTHF is very high near the critical point of

TABLE 5.2
Mark–Houwink Equations for PTHF, $[\eta] = KM^a$

Solvent	T, °C	$K \times 10^3$ ml/g	a	$M \times 10^{-4}$	Method[a]	Ref.
Benzene	20	25.9	0.82	0.075–1.28	VPO[b]	14
	20	25	0.82	"low"	OS	15
	25	26.2	0.80	0.08–2.5	EG	16
	25	29.8	0.79	—	OS	17
	30	131	0.60	3.5–110	LS	5
	30	314	~0.5	—	KSF	5
Toluene	28	25.1	0.78	3.2–11.8[c]	OS	12
	30	27.3	0.81	0.12–0.94	VPO[b]	14
	30	23.6	0.777	0.18–1.26	OS	18
n-Butanol	20–28	41.7	0.68	3.7–20.5[c]	OS	6
Ethyl acetate	30	42.2	0.65	3.5–110	LS	5
	30	196	~0.5		KSF	5
Cyclohexane	30	176	0.54	3.5–110	LS	5
	30	243	~0.5	—	KSF	5
Isopropyl alcohol	44.6	231	0.50	3.8–21.0[d]	OS	6
Methyl ethyl ketone	25	210	0.5	1.7–150	LS	10
Ethyl acetate/ n-hexane:						
22.7/77.3 wt. %	31.8	343	0.45	3.5–110	LS	5
22.3/77.7 wt. %	30.4	267	0.50	3.7–21.0[d]	OS	6
22.7/77.3 wt. %	31.8	218	~0.5	—	KSF	5
Several solvents	25	202	0.5	77	OS	19, 20

[a]Abbreviations: OS, osmometry; VPO, vapor-pressure osmometry; LS, light scattering; EG, end-group analysis; KSF, method of Kurata *et al.*[27,28]
[b]Unfractionated.
[c]Ten fractions.
[d]Fractionated.

methanol–cyclohexane mixtures. Turbidimetric measurements indicated that a small quantity of polymer has a large unfavorable effect on the miscibility of the two solvents. The dimensions of the PTHF in solution are abnormally high even when compared to those in pure cyclohexane.

Slope constants Numerous determinations of viscosity slope constants for solutions of PTHF have been made.[5–7,22] The two equations most commonly used are:

$$\eta_{sp}/C = [\eta] + k_H[\eta]^2C = [\eta](1 + k_H[\eta]C)$$

and

$$\eta_{sp}/C = [\eta] + k_{SB}[\eta]\eta_{sp} = [\eta](1 + k_{SB}\eta_{sp}),$$

where η_{sp} is the specific viscosity; C is the concentration; k_H is the Huggins constant;[23,24] and k_{SB} is the Schulz–Blaschke constant.[25] For PTHF the values of k_{SB} lay in the range 0.22–0.61, while the Huggins constants were 0.29–0.96 and even greater than 1.1. The highest slope constants were obtained under theta conditions. Elias and Adank,[22] who obtained the values of $k_H > 1.1$, proposed a solvent-induced helical structure involving weak induction forces between the protons of the methylene groups and the main-chain oxygen atoms. According to Howard *et al.*,[7] there is not enough evidence to support this hypothesis, especially since there are no clear chemical and structural differences between the solvents that promote the effect and those that do not.

Elias and Adank[22] observed a linear relationship between k_H and $[\eta]$. They determined values of $[\eta]$ and k_H in a wide variety of solvents, including butanone, benzyl alcohol, chlorobenzene, benzene, toluene, carbon tetrachloride, tetrahydrofuran, dioxane, 3-methyl-2-heptanone, 2-heptanone, 2-pentanone, butanone and methyl acetate. In contrast Ali and Huglin[12] found a nonlinear relationship. For the highest molecular weights, k_H was in the range from 0.47 to 0.41; for the lowest molecular weights, k_H varied from 0.38 to 0.33.

5.4.2 Effect of molecular weight

Some effects of molecular weight on intrinsic viscosity are given in Table 5.3.[5] Kurata *et al.*[5] used unfractionated PTHF samples. They found that, for one polymer of 21,000 molecular weight, the ratio of weight-average to number-average molecular weight was 1.04, and assumed that the same was true for all the polymers studied.

TABLE 5.3
Effect of Solvent on Intrinsic Viscosity, $[\eta]$, dl/g, of PTHF at 30°C[5]

$M_w \times 10^{-4}$	Ethyl acetate	Benzene	Cyclohexane	Theta solvent[a]
3.46	0.402	0.700	0.473	0.367
4.71	0.514	0.912	0.590	0.440
20.2	1.20	1.836	1.125	0.828
40.3	2.07	3.31	1.900	1.115
68.8	2.80	4.25	2.75	1.352
103.0	3.21	5.38	—	1.590

[a]Ethyl acetate/*n*-hexane (22.7 wt.%/77.3 wt.%) at theta-temperature of 31.8°C.

Huglin and Whitehurst[15] used low-molecular-weight hydroxy-terminated PTHF and other low-molecular-weight polymers to test the proposition that low-molecular-weight fractions in good solvents behave as if they were under theta conditions. The general conclusion was that the proposition is invalid.

5.4.3 Effect of temperature

The effect of temperature on the intrinsic viscosity of PTHF in 21 different solvents was reported by Evans *et al.*[26] Many polymers exhibit linear plots of $\ln [\eta]_T$ against T. $[\eta]_T$ is the intrinsic viscosity at constant temperature T. This was almost true for PTHF dissolved in methyl ethyl ketone, *n*-amyl-*n*-butyrate, *iso*-butyl-*n*-butyrate, *iso*-butyl acetate and *n*-propyl formate. However, as shown in Figure 5.1, it was not true for the 16 other solvents, which varied in behavior. Some of the plots increased, some decreased and others showed a maximum as a function of temperature.

5.4.4 Effect of copolymeric units

Intrinsic-viscosity–molecular-weight relationships for PTHF usually cannot be applied to copolymers containing PTHF units. For any given molecular weight, the intrinsic viscosity may be lower, higher or similar depending on the nature of the

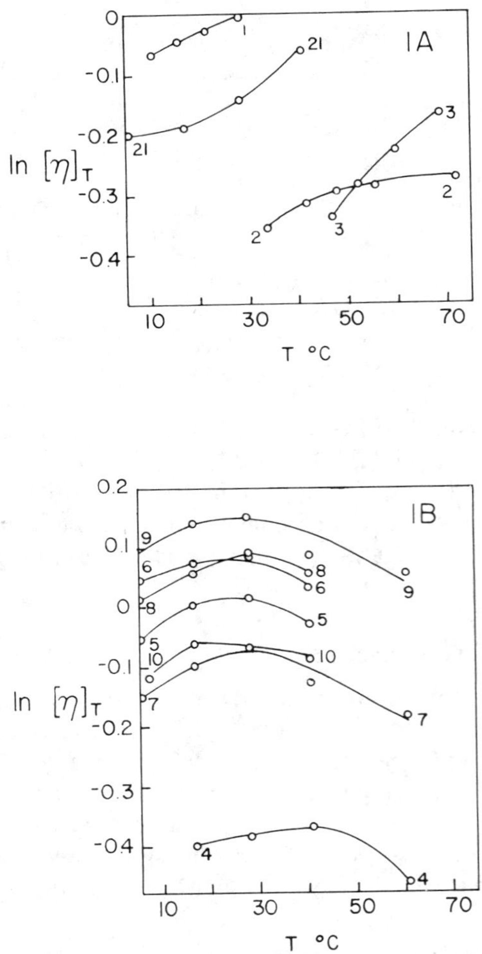

FIGURE 5.1 Variation of ln $[\eta]_T$ with temperature for solutions of PTHF in different solvents: 1) *n*-butanol; 2) diethyl malonate; 3) *iso*-propyl alcohol; 4) ethyl lactate; 5) ethyl acetate; 6) *n*-butyl acetate; 7) *n*-butyl formate; 8) *n*-propyl acetate; 9) *n*-amyl acetate; 10) *sec*-butyl acetate; 11) *iso*-propyl acetate; 12) ethyl *n*-butyrate; 13) chlorobenzene; 14) *n*-butyl propionate; 15) *iso*-amyl acetate; 16) methyl ethyl ketone; 17) *n*-amyl *n*-butyrate; 18) *iso*-butyl *n*-butyrate; 19) *iso*-butyl acetate; 20) *n*-propyl formate; 21) ethyl formate. The curves are divided into increasing plots (A), plots with a maximum (B) and decreasing plots (C).[26]

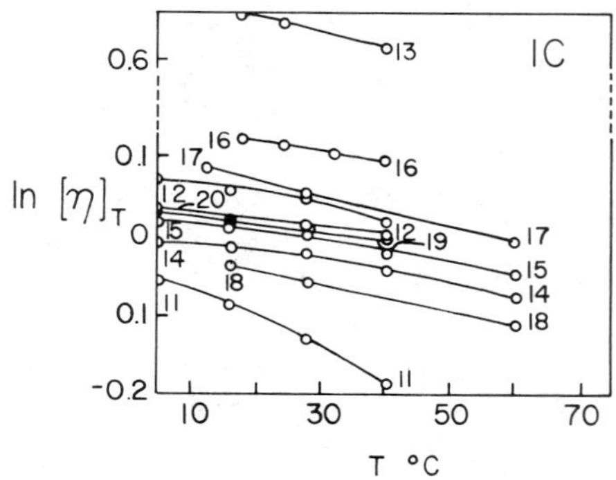

FIGURE 5.1 (*continued*).

copolymeric unit, its interaction with the selected solvent and the sequence distribution of the units comprising the copolymer.

5.5 UNPERTURBED DIMENSIONS

The dimensions of macromolecular chains in solution are described by three different quantities: $\langle r^2\rangle_0$, $\langle r^2\rangle$ and $\langle r^2\rangle_{0f}$.

The root-mean-square unperturbed end-to-end distance, $\langle r^2\rangle_0$, is measured under ideal or theta conditions. It can also be obtained by extrapolation of data obtained in good solvents. It is a measure of short-range interactions. One way to obtain $\langle r^2\rangle_0$ is to plot $[\eta]/M^{1/2}$ against $M^{1/2}$ according to the theory of Kurata *et al.*[27,28]:

$$[\eta] = KM^{1/2} + 0.51\phi_0 BM$$

$$K = \phi_0 A^3 = \phi\langle r^2\rangle_0/M^{3/2},$$

where A and B are parameters which characterize the conformational properties of a polymer chain in a solvent; and ϕ_0 is the Flory universal constant, whose theoretical value lies between 1.40×10^{21} and 2.87×10^{21} mol^{-1}.

The root-mean-square end-to-end distance, $\langle r^2 \rangle$, can be obtained from light-scattering data, and $\langle r^2 \rangle_{0f}$, the theoretical value of dimensions of a freely rotating chain, is calculated from known bond-angle and bond-length considerations.

Although separate values of $\langle r^2 \rangle_0$ and $\langle r^2 \rangle$ are reported, they are very often combined in the ratio $\alpha = (\langle r^2 \rangle / \langle r^2 \rangle_0)^{1/2}$, which is called the expansion factor; α represents the effect of "long-range" interactions, and is a measure of the osmotic swelling of a polymer chain by polymer–solvent interactions.

Steric factors are commonly described by a conformational factor, $\sigma = (\langle r^2 \rangle_0 / \langle r^2 \rangle_{0f})^{1/2}$.

Studies leading to values of all these parameters have been carried out repeatedly for PTHF.[3,5,6,12,23,26,29,30,31]

Some experimental values of α are given in Table 5.4. The differences in results reported by the two groups of investigators are probably due to differences in molecular weights of the PTHF samples used by the groups. The large values of α in chlorobenzene and toluene indicate considerable osmotic swelling in good solvents like toluene and chlorobenzene. The expansion is less in methyl ethyl ketone but still more than in theta solvents, where $\alpha = 1.00$.

The limiting value calculated for $(\langle r^2 \rangle_{0f}/M)^{1/2}$ is

TABLE 5.4
Values of the Expansion Factor, α_η, for PTHF

Solvent	T, °C	α_η[a]	Ref.
Chlorobenzene	25	1.27	19, 20
	25	1.39	6
Toluene	25	1.24	19, 20
	28	1.34	6
Methyl ethyl ketone	25	1.07	19, 20
	25	1.16	6
Theta solvents	—	1.00[b]	19, 20
	—	1.00	6

[a]Obtained from viscosity measurements.
[b]From mean value of $[\eta]_\theta$ obtained by various methods.

0.5583 Å $mol^{1/2}\,g^{-1/2}$, while that for $(\langle r^2\rangle_{0f}/N)^{1/2}$ is 2.120 Å.[31] Here M is the molecular weight and N is the total number of bonds in the chain.

The factor σ is 1.85 in benzene, 1.70 in cyclohexane, 1.58 in ethyl acetate and 1.64 in the theta solvent ethyl acetate/n-hexane (22.7/77.3 wt.%).[5] In addition, σ is 1.67 in isopropanol, 1.69 in diethyl malonate and 1.75 in ethyl acetate/n-hexane (22.7/77.3 wt.%).[6] The average of these values, 1.67–1.70, lies between those for polyethylene oxide and polyethylene. The effect of benzene on chain conformation is notable.

May and Wetton[4] obtained another indication that benzene leads to an expansion of polymer coils. Their viscosity measurements indicated that $\langle r^2\rangle_0^{1/2}$ in benzene was 1.34 times as great as that in the mixed theta solvent consisting of heptane/cyclohexane (39.2/60.8 wt.%).

Evans *et al.*[26,3] examined the variation of the root-mean-square unperturbed end-to-end distance as a function of temperature, $d\ln\langle r^2\rangle_0/dT$, and compared the results with predictions from several excluded-volume theories. The analysis was not wholly successful because, as shown in Figure 5.1, the relation of intrinsic viscosity with temperature varies irregularly with solvent. For the five solvents where a linear relationship between viscosity and temperature was obtained, the results of Evans *et al.*[26] plus those taken from other literature indicate that $d\ln\langle r^2\rangle_0/dT$ lies in the range from -0.6×10^{-3} to $-3\times10^{-3}\,deg^{-1}$.

Bak *et al.*[30] have also measured $d\ln\langle r^2\rangle_0/dT$. They obtained their values from force–temperature measurements on cross-linked networks. Their average value, $-1.33\times10^{-3}\,deg^{-1}$ in the vicinity of 60°C is within the range quoted above.

5.6 MOLECULAR-WEIGHT DETERMINATION

5.6.1 Number-average molecular weight

Number-average molecular weights, $\bar{M}_n$, of PTHF have been determined by standard techniques. There is nothing unusual about the behavior of PTHF in these procedures and they can be carried out in the standard way.

For polymers of low molecular weight $\bar{M}_n$ has been calculated primarily from data obtained by end-group analysis. Depending on the nature of the end group, the methods used have included titration,[14,18,32] NMR,[33,34] IR,[35–37] UV,[35,36] elemental analysis,[35,36,38] liquid-scintillation counting of ^{14}C-labelled end groups[39] and fluorescence spectroscopy.[38] The variety of end groups that can be introduced and some methods of analysis have already been discussed in the termination section of Chapter 3. Ebulliometry[11] and vapor-pressure osmometry in benzene[14,39] have also been used.

For polymers with $\bar{M}_n$ values of 15,000 and above, osmometry is most frequently used. Toluene and benzene are the most common solvents,[11,12,14,19,20,40] although other solvents have also been used to determine $\bar{M}_n$ and the variation of the second virial coefficient, A_2, with temperature. Typical results for the variation of A_2 in toluene with temperature for a PTHF sample with $\bar{M}_n$ of 118,000 are shown in Figure 5.2. In theta solvents, by definition, the value of A_2 is zero, and this is one method by which solvents for PTHF have been shown to be theta solvents.

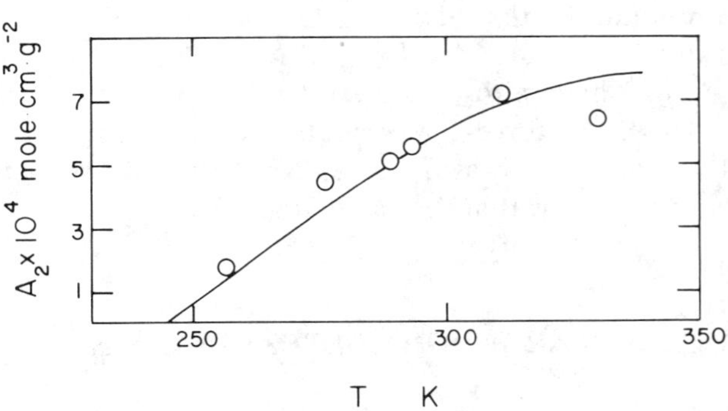

FIGURE 5.2 Variation with temperature of second virial coefficients, A_2, for PTHF in toluene.[12] The data were obtained by osmometry.

5.6.2 Weight-average molecular weight

Weight-average molecular weights of PTHF, $\bar{M}_w$, can be determined readily by light scattering.[3,5,6,10,19,20] A variety of solvents and all the usual instruments can be, and have been, used. Specific-refractive-index increments, dn/dc, and refractive indices, n_D, for typical solvents are given in Table 5.5.

TABLE 5.5
Refractive Index and Specific Refractive-index increment (dn/dc) for PTHF in Various Solvents

Solvent	n_D^{25}	dn/dc, ml/g	λ, mμ	Ref.
Chlorobenzene	1.5247_9	0.070[a]	436	10
Isopropyl acetate	1.3830[b]	0.098[b]	436	3
Isopropanol	1.3776	0.108[c]	436	6
Ethyl acetate	1.3698	0.113_6	436	5
Ethyl acetate/n-hexane (22.7/77.3 wt.%)	1.3684	0.113_9	436	5
3-Methyl-2-heptanone	1.4143	0.056	546	19, 20
Tetrahydrofuran	1.4036	0.064	546	19, 20
2-Pentanone	1.3878	0.084	546	19, 20
Methyl ethyl ketone	1.3807_1	0.095	546.1	10
2-Butanone	1.3700	0.091	546	19, 20

[a]dn/dc values for the chlorobenzene/n-octane theta solvents listed in Table 5.1 were small and ranged from 0.046 to 0.078 ml/g.[11]
[b]At 22.5°C.
[c]At 46°C.

Some typical values of the second virial coefficient as a function of molecular weight are shown in Figure 5.3 for samples of PTHF dissolved in ethyl acetate at 30°C.[5]

Values of the z-average of the root-mean-square radius of gyration of PTHF, $\langle S^2\rangle^{1/2}$, in ethyl acetate at 30°C are in the range $(417–528)\times 10^{-8}$ cm for the polymers in Figure 5.3.

Evans and Huglin[3] have used light scattering to determine the temperature dependence of the unperturbed radius of gyration of PTHF dissolved in chlorobenzene/n-octane mixtures. Their results are shown in Figure 5.4. The unexpected

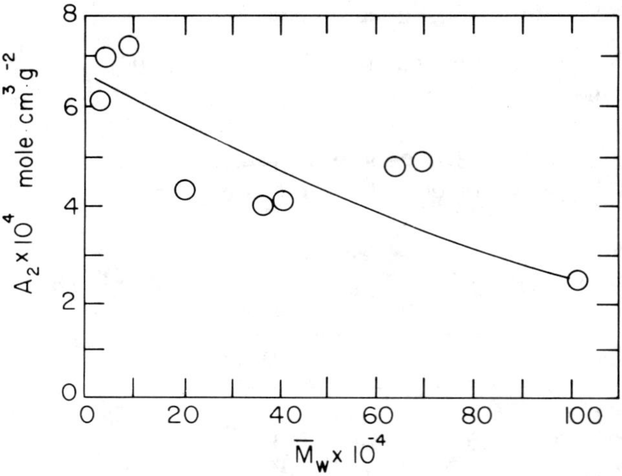

FIGURE 5.3 Variation with molecular weight of second virial coefficients, A_2, for PTHF in ethyl acetate.[5] The data were obtained by light scattering.

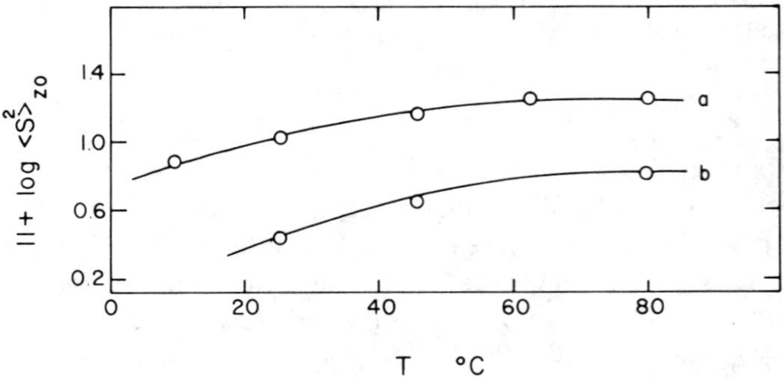

FIGURE 5.4 Variation of unperturbed dimensions (in cm^2) with temperature. The values were obtained from light scattering in mixtures of chlorobenzene and *n*-octane. Values are given for two different molecular weights: (a) $M = 1.44 \times 10^6$; (b) $M = 0.56 \times 10^6$.[3] "Reprinted with permission from European Polymer Journal, *6*, 1161 (1970), Pergamon Press, Ltd."

increase with temperature that they found conflicts with results from viscometry and with theoretical predictions.[3] They therefore conclude that, in agreement with Yamakawa's proposition, coil sizes measured by light scattering are likely to be considerably in error when preferential absorption of a good solvent like chlorobenzene prevails.[3]

5.6.3 Molecular-weight distributions

Studies of the molecular-weight distribution of PTHF and of its changes during a "living" polymerization of PTHF have provided new experimental information that was not readily obtained with "living" anionic polymerizations. But there is good reason to expect that yet more definitive information will be obtained in the future, as synthetic chemists become more confident about their ability to control the polymerization variables that affect molecular-weight distribution. Numerous claims at control have been made but the results must be judged critically in the light of the total information available about THF polymerizations. It is not uncommon to find that a result obtained in one year is modified in the next year by the same group of authors.[41–43]

THF polymerization seems especially well suited as a model system for testing theoretical treatments of the effect of various reactions occurring during polymerization on molecular-weight distributions. Among the attractive features are the following: it is a "living" polymerization, with significant rates of both propagation and depropagation; chain transfer with polymer oxygen is possible, but there are indications that it can be eliminated; efficient initiators where $k_i \gg k_p$ or $k_i < k_p$ are known; and finally, very efficient chain-transfer agents are known.

Many features of the molecular-weight distributions obtained during THF polymerizations are well-enough documented and often-enough reproduced to be accepted.

Methods of determination Molecular-weight distributions of polymers obtained in THF polymerizations have been determined from $\bar{M}_w/\bar{M}_n$ ratios derived from light scattering and osmometry,[11] turbidimetric titration,[11] sedimentation velocity in a theta solvent[5,44] and GPC.[18,37,41,42,45] Note that, although dn/dc is small for THF, as shown in Table 5.5, it is large enough for THF to be used as a solvent for measurements of the molecular-weight distribution of PTHF by means of a gel-permeation chromatograph having a differential refractometer as detector.

Theoretical predictions Brown and Szwarc[46] originally pointed out that a normal ("most probable") distribution of molecular weights will result from a "living" polymerization. However, if the depropagation reaction is slow, and initiation is rapid, a narrow (Poisson) distribution will form initially. Miyake and Stockmayer[47] have also investigated the mathematical problem of a reversible "living" polymer system without transfer or termination reactions. They also concluded that initially a Poisson distribution will form, and that this distribution will change with time until eventually the "most probable" distribution where $\bar{M}_w/\bar{M}_n$ is two will be formed. Hermans[2] has calculated the resulting chain-length distribution in a polymer in which chain ends react at random with all monomer units. His calculations show that ultimately a "most probable" distribution will be obtained, and that the "most probable" distribution is achieved more rapidly as the molecular weight of the polymer increases. Höcker[48] has calculated that, using a bifunctional initiator, the distribution obtained will be composed of a single distribution and a coupled one. The maximum value of $\bar{M}_w/\bar{M}_n$ will never exceed 1.4 even for a ratio $k_p/k_i = 10^6$. Taganov *et al.*[49] have investigated the molecular-weight distribution of polymers in the cationic polymerization of cyclic ethers. They conclude that changes in molecular-weight distribution can be used as a means of determining the reactions that are taking place, and that rate constants for the reactions can be determined from the experimental results.

Experimental results For polystyrene polymerized anionically, the initial Poisson distribution is produced in a matter of seconds, but the rate of depropagation is so slow that final equilibrium is not predicted for 80 year or so.[47]

Polystyrene of narrow molecular-weight distribution is readily prepared. The depropagation reaction occurs much more readily in PTHF polymerizations. Calculations according to the theory of Miyake and Stockmayer[47] show that, in the absence of termination and transfer reactions, a normal distribution should be obtained in about 20 days.[58] The preparation of PTHF with very narrow molecular-weight distribution in the early stages of the polymerization has been reported repeatedly.[5,18,39,41,45,50,51,58] PTHF samples with $\bar{M}_w/\bar{M}_n <$ 1.1 at equilibrium conversion and polymerization temperatures of 0°C and above are rarely obtained.[8,10,39,45,52] Polymers having a normal distribution or even higher distributions when $k_i \ll k_p$ are common at these higher conversions.

There are four primary causes for these broader molecular-weight distributions. The first is inefficient initiation with $k_i \ll k_p$ and initiation that continues throughout polymerization. Such initiation can also be accompanied by transfer reactions, as in the case of $SbCl_5$, where $\bar{M}_w/\bar{M}_n$ ratios as high as 2.8 and multimodal distributions have been reported.[10] The second is the relatively rapid depropagation rate. Even with very efficient initiation, in time the molecular weight distribution would become broader. However, this cause is not sufficient to explain the experimental results. PTHF polymerizations are rarely, if ever, continued for the 20 days that are predicted on the basis of the known rates of propagation and depropagation to be required to reach a normal distribution. Thus a third cause has been invoked: transfer to polymer oxygen to form an acyclic oxonium ion, followed by further reaction with THF.[58] Finally, the effect of end groups needs to be considered. Generally speaking, in anionic polymerization the head group is an unreactive alkyl group. In THF polymerizations, the head group may be a reactive group such as a hydroxyl group. In such cases unexpectedly high molecular weights and polymers with broad molecular-weight distributions of molecular weights can be obtained[52] (see Section 4.2.7).

PTHF samples with very narrow distributions even at high conversions have been reported in experiments using high-vacuum techniques and other special procedures such as initiation above the ceiling temperature at 95°C.[8,44] The final isolation and purification of the polymers could have led to loss of species of very low molecular weight. The work is significant because at least in one case[8] it seems to indicate that, at −10°C, transfer to polymer oxygen can be virtually eliminated and, provided initiation is fast, a narrow molecular-weight distribution will be obtained. Further work by Richards and coworkers lends support to this suggestion.[41,42] But it seems clear, especially in view of the difficulties experienced by Richards and coworkers in reproducing their original results,[43] that further synthetic work is required before definitive theoretical studies can be made.

Effect of added monomeric transfer agent It is not entirely clear what effect adding transfer agent will have on the $\bar{M}_w/\bar{M}_n$ ratio for the resulting PTHF. In one study, Rozenberg *et al.*[11] found that adding water as transfer agent led to broadening of the molecular-weight distribution as the conversion increased. Some data are shown in Table 5.6. According to these results,

TABLE 5.6
Influence of the Addition of Water on the Molecular-weight Distribution of PTHF[11] at Equilibrium Conversion[a]

$[H_2O]:[I]$	$\bar{M}_w \times 10^{-4}$	$\bar{M}_n \times 10^{-4}$	$\bar{M}_w/\bar{M}_n$
0	30	15.6	2.00
0.04	19	5.7	3.34
	23	7.0	3.28[b]
0.33	14	2.0	7.00
	18.1	3.1	5.85[b]

[a]The initiator was $[Et_3O^{\oplus}BF_4^{\ominus}] = 0.02$ M. Polymerization was in bulk THF at 20°C.

[b]Calculated from integral molecular-weight-distribution curves.

TABLE 5.7
Trimethylorthoformate Added at Equilibrium[45] to a Tetrahydrofuran Polymerization[a]

TMOF/I[b]	Hours[c] Before	Hours[c] After	$[\eta]$[d]	$\bar{M}_w \times 10^{-4}$	$\bar{M}_n \times 10^{-4}$	$\bar{M}_w/\bar{M}_n$
0	8	0	1.00	6.19	2.92	2.12
0.5	8	8	0.67	2.98	1.74	1.70
0	8	8	1.00	6.50	3.10	2.09

[a]Polymerizations were carried out in 37.5 vol.%, methylene chloride solutions at 30°C in sealed ampoules. The initiator was triethyloxonium hexafluorophosphate at a concentration of 3.25×10^{-3} M. Standard high vacuum techniques were used throughout. It was previously shown that eight hours are sufficient to achieve equilibrium conversion under these conditions.

[b]TMOF/I is the molar ratio of trimethylorthoformate to initiator.

[c]These two columns give the time the polymerization was allowed to proceed before the addition of the trimethylorthoformate and then the corresponding number of hours after the addition.

[d]Viscosities were run in ethyl acetate at 30°C.

[e]GPCs were run in tetrahydrofuran at 40°C.

the effect of added monomeric transfer agent is to broaden the molecular-weight distribution. However, in assessing these results, it must be remembered that hydroxyl end groups are produced by the reaction of THF tertiary oxonium ions with water. In 1965, the coupling and other complicating reactions that can occur in the presence of hydroxyl groups[52,53] were not fully appreciated. It may be these other reactions, rather than transfer reactions alone, which account for the broad molecular-weight distributions obtained in the presence of water. In addition, it is now known that termination reactions occur with $BF_4^{\ominus}$ counterions above 0°C, and these reactions would also result in a broadened molecular-weight distribution.

In contrast to the results with added water, Dreyfuss and Dreyfuss[45] have reported the data shown in Table 5.7. In this case the transfer agent, trimethylorthoformate, was added after equilibrium conversion was attained. Control polymerizations without added transfer agent had normal molecular-weight distributions with $\bar{M}_w/\bar{M}_n \simeq 2$. When trimethylorthoformate was added, the overall molecular weight was reduced and the molecular-weight distribution was reduced substantially. In fact, Dreyfuss and Dreyfuss had shown earlier that, in the presence of a large amount of trimethylorthoformate, after 14 days only oligomers having the structure $CH_3O(CH_2CH_2CH_2CH_2O)_nCH_3$ are produced and, of these, 43.3% had $n = 1$, 21.2% had $n = 2$ and 35.3% had $n > 2$, with an average n of 3.5.[40] This extreme case gives $\bar{M}_w/\bar{M}_n \simeq 1.2$. Even if the value is somewhat higher, as it will be if real values rather than the average $n = 3.5$ is used for calculating $\bar{M}_w$ when $n > 2$, the ratio is clearly less than two. In these experiments, the counterion $PF_6^{\ominus}$ is reasonably stable under the conditions used, initiation was complete and conditions of equilibrium conversion as well as equilibrium molecular-weight distribution were attained before transfer agent was added. There are good grounds to doubt that scission of the chains by added transfer agent necessarily leads to a normal distribution.

5.7 SOLUBILITY PARAMETER AND COHESIVE ENERGY DENSITY

Huglin and Pass[54] have measured the solubility parameter, δ_p, of PTHF in several ways. Using 17 simple esters with solubility parameters ranging from 7.8 to 10.0 $(cal/cm^3)^{1/2}$, they obtained the results given in Table 5.8 for a PTHF polymer which was prepared using PF_5 as initiator and sometimes radiation-cross-linked. The mean of nine values was 8.55 $(cal/cm^3)^{1/2}$. From this they calculated a value of 5.17 kcal mole^{-1} for the cohesive energy density.

Huglin and Pass have examined the effect of solvent power on the Huggins slope constant k_H and have related it to δ_s, the solubility parameter of the solvent.[54] They find that, normally, k_H is 0.3–0.4 in good solvents with the minimum k_H at $\delta_s = 8.4$–8.7. Higher values of k_H are exhibited by poor solvents. Thus, in both *sec*-butyl acetate ($\delta_s = 8.2$) and ethyl acetate ($\delta_s = 9.05$), k_H is 0.64.

TABLE 5.8
Solubility Parameters for PTHF[54]

δp, $(cal/cm^3)^{1/2}$	Method
8.6 ± 0.1	Direct swelling measurements of crosslinked PTHF
8.6 ± 0.1	Swelling measurements allowing for the molar volume of the solvent
8.6 ± 0.1	Intrinsic viscosity of linear polymer solutions
8.6 ± 0.1	Intrinsic viscosity allowing for molar volume of the solvent
	Swelling in conjunction with stress–strain measurements:
8.44 ± 0.05	from slope
8.54 ± 0.05	from intercept
	Incorporating χ derived from Flory theta condition:
8.48 ± 0.05	from slope
8.54 ± 0.05	from intercept
8.52	Calculation from molar attraction coefficients

5.8 DIPOLE MOMENT AND DIELECTRIC-CONSTANT MEASUREMENTS

Several groups have examined the dielectric properties of PTHF in solution.[30,55,56] From these, they calculated the dipole-moment ratio $\langle \mu^2 \rangle/(Nm^2)$, where $\langle \mu^2 \rangle$ is the mean dipole moment; m^2 is the average of the squares of the bond moments; and N is the number of skeletal bonds in the chain. It was assumed that the skeletal bonds are freely jointed. Therefore, the dipole-moment ratio represents the observed mean-square dipole moment of the chain relative to the mean-square moment of the freely jointed chain. The values found are tabulated in Table 5.9. They vary from 0.586 to 0.654 in the temperature range 20 to 60°C and are essentially independent of molecular weight. (The values for the lower molecular weight are slightly higher because of end-group effects.) The values are reported to be in agreement with predictions from isomeric rotational theory.[56]

TABLE 5.9
Dielectric Results and Dipole-moment Ratios for PTHF

$M \times 10^{-3}$	T,°C	Solvent	$\frac{d}{dw}(\epsilon - \epsilon_1)$[a]	$\langle \mu^2 \rangle/(Nm^2)$[b]	Ref.
500[c]	20	None[d]	—	0.53–0.64[e]	55
408[f]	20	C_6H_6	1.70	0.586	30
	30		1.68	0.606	
	40		1.66	0.625	
	50		1.64	0.640	
	60		1.61	0.654	
2.5[g]	20	C_6H_6	1.90	0.603	56
	30		1.86	0.621	
	40		1.80	0.632	
	50		1.74	0.642	
	60		1.68	0.647	

[a] ϵ and ϵ_1 are static dielectric constants of solution and solvent, respectively; w is the weight fraction of polymer.

[b] See definition in text.

[c] $\bar{M}_w$.

[d] Undiluted, quenched sample, described as wholly amorphous.

[e] The wide range results from the difficulty in estimating a value of the limiting high-frequency dielectric constant of the bulk PTHF.

[f] $\bar{M}_v$.

[g] $\bar{M}_n$.

5.9 NEUTRON SCATTERING

High-resolution quasielastic neutron scattering of PTHF in dilute CS_2 solution at 10°C has been studied.[57] The B. H. Zimm model for polymers in dilute solution gave the best fit of three models tested. The results did not depend on molecular weight, but deviations occurred at temperatures different from 10°C.[57]

REFERENCES

1. J. Brandrup and E. H. Immergut (eds.), "Polymer Handbook", Wiley, New York, NY, p. IV-168 (1975).

2. J. J. Hermans, *J. Polym. Sci.*, **C12**, 345 (1966).
3. J. M. Evans and M. B. Huglin, *Europ. Polym. J.*, **6**, 1161 (1970).
4. R. W. May and R. E. Wetton, Private communication.
5. M. Kurata, H. Utiyama and K. Kamada, *Makromol. Chem.*, **88**, 281 (1965).
6. J. M. Evans and M. B. Huglin, *Makromol. Chem.*, **127**, 141 (1969).
7. N. Howard, M. B. Huglin and R. W. Richards, *J. Appl. Polym. Sci.*, **16**, 1525 (1972).
8. T. G. Croucher and R. E. Wetton, *Polymer*, **17**, 205 (1976).
9. F. J. Burgess, A. V. Cunliffe, D. H. Richards and D. Thompson, *Polymer*, **19**, 334 (1978).
10. N. V. Makletsova, I. V. Epel'baum, B. A. Rozenberg and E. B. Lyudvig, *Vysokomol. Soyed.*, **7**, 70 (1965) (in Russian). Translated in *Polym. Sci. USSR*, **7**, 73 (1965).
11. B. A. Rozenberg, N. V. Makletsova, I. V. Epel'baum, E. B. Lyudvig and S. S. Medvedev, *Vysokomol. Soyed.*, **7**, 1051 (1965) (in Russian). Translated in *Polym. Sci. USSR*, **7**, 1163 (1965).
12. S. M. Ali and M. B. Huglin, *Makromol. Chem.*, **84**, 117 (1965).
13. A. Davis and J. H. Golden, *Makromol. Chem.*, **81**, 38 (1965).
14. D. Sims, *Makromol. Chem.*, **98**, 235 (1966).
15. M. B. Huglin and D. H. Whitehurst, *J. Appl. Polym. Sci.*, **12**, 1889 (1968).
16. Y. Hachihama and T. Shono, *Technol. Rep. Osaka Univ.*, **9**, 337 (1959).
17. R. M. Bell, C. Fitzsimmons and A. Ledwith, *Polymer*, **6**, 661 (1965).
18. E. A. Ofstead, *ACS Polym. Preprints*, **6**(2), 674 (1965).
19. G. Adank, Thesis No. 3703, ETH Zürich, Clearinghouse Publications No. 66-36762, Springfield, VA. 22151, USA (1965).
20. H.-G. Elias and G. Adank, *Makromol. Chem.*, **103**, 230 (1967).
21. A. Dondos and Y. Izumi, *Makromol. Chem.*, **181**, 701 (1980).
22. H.-G. Elias and G. Adank, *Makromol. Chem.*, **69**, 241 (1963).
23. H.-G. Elias and O. Etter, *Makromol. Chem.*, **66**, 56 (1963).
24. F. Ibrahim and H.-G. Elias, *Makromol. Chem.*, **76**, 1 (1964).
25. G. V. Schulz and F. J. Blaschke, *J. Prakt. Chem.*, **158**, 130 (1941).
26. J. M. Evans, M. B. Huglin and R. F. T. Stepto, *Makromol. Chem.*, **146**, 91 (1971).
27. M. Kurata and W. H. Stockmayer, *Adv. Polym. Sci.*, **3**, 196 (1963).
28. W. H. Stockmayer and M. Fixman, *J. Polym. Sci.*, **C1**, 137 (1963).
29. M. B. Huglin and R. F. T. Stepto, *Makromol. Chem.*, **132**, 225 (1970).
30. K. Bak, G. Elefante and J. E. Mark, *J. Phys. Chem.*, **71**, 4007 (1967).
31. J. M. Evans, M. B. Huglin and R. F. T. Stepto, *Makromol. Chem.*, **125**, 59 (1969).
32. Y. Hachihama and T. Shono, *Technol. Rep. Osaka Univ.*, **9**(361), 299 (1959).
33. A. Hase and T. Hase, *Analyst*, **97**, 998 (1972).
34. Y. Eckstein, D. P. Lee, R. P. Quirk and P. Dreyfuss, *J. Polym. Sci.: Poly. Chem. Ed.*, **18**, 2021 (1980).

35. D. Sims, *Makromol. Chem.*, **98**, 235 (1966).
36. J. Lehmann, L. Reibel, E. Franta, W. Dobrogoszcz and S. Penczek, "Abstracts to First International Symposium on Polymerizations of Hetrocycles (Ring-opening)", Warsaw-Jablonna, Poland, p. 141 (June 23–25, 1975).
37. J. Lehmann, PhD Thesis, University Louis Pasteur of Strasbourg (June 11, 1976).
38. Y. Eckstein and P. Dreyfuss, *Anal. Chem.*, **52**, 537 (1980).
39. D. Vofsi and A. V. Tobolsky, *J. Polym. Sci.*, **A3**, 3261 (1965).
40. M. P. Dreyfuss and P. Dreyfuss, *J. Polym. Sci.: A1*, **4**, 2179 (1966).
41. D. H. Richards, S. B. Kingston and T. Souel, *Polymer*, **19**, 68 (1978).
42. D. H. Richards, S. B. Kingston and T. Souel, *Polymer*, **19**, 806 (1978).
43. D. H. Richards and D. Thompson, *Polymer*, **20**, 1439 (1979).
44. T. Fujimoto, M. Kawahashi, M. Nagasawa and A. Takahashi, *Polym. J.*, **11**, 193 (1979).
45. M. P. Dreyfuss and P. Dreyfuss, *ACS Polym. Preprints*, **11**, 203 (1970).
46. W. B. Brown and M. Szwarc, *Trans. Faraday Soc.*, **54**, 416 (1958).
47. A. Miyake and W. H. Stockmayer, *Makromol. Chem.*, **88**, 90 (1965).
48. H. Höcker, *Makromol. Chem.*, **157**, 187 (1972).
49. N. G. Taganov, G. V. Korovina and S. G. Entelis, *Vysokomol. Soyed.*, **A20**, 1393 (1978) (in Russian). Translated in *Polym. Sci. USSR*, **20**, 1569 (1978).
50. E. Franta, L. Reibel, J. Lehmann and S. Penczek, *J. Polym. Sci.: Polym. Symp.*, **56**, 139 (1976).
51. F. Afshar-Taromi, M. Scheer, P. Rempp and E. Franta, *Makromol. Chem.*, **179**, 849 (1978).
52. G. Pruckmayr and T. K. Wu, *Macromolecules*, **11**, 662 (1978).
53. G. Pruckmayr and T. K. Wu, *Macromolecules*, **11**, 265 (1978).
54. M. B. Huglin and D. J. Pass, *J. Appl. Polym. Sci.*, **12**, 473 (1968).
55. R. E. Wetton and G. Williams, *Trans. Faraday Soc.*, **61**, 2132 (1965).
56. E. Riande, *Makromol. Chem.*, **178**, 2001 (1977).
57. J. S. Higgins, G. Allen, R. E. Ghosh, W. S. Howells and B. Farnoux, *Chem. Phys. Lett.*, **49**(2), 197 (1977).
58. P. Dreyfuss and M. P. Dreyfuss, *Adv. Chem. Series*, **91**, 335 (1969).

CHAPTER 6

Bulk Properties and Some Applications

6.1 INTRODUCTION

Solution properties of PTHF were discussed in Chapter 5. This chapter is concerned primarily with properties of the bulk polymer, in general. Detailed discussion of the properties of typical low-molecular-weight PTHF glycols and diisocyanates is in Chapter 7, which deals with the synthesis, properties and uses of polyurethanes and polyesters prepared from PTHF.

6.2 DENSITY

Crystalline and amorphous densities of PTHF taken from a variety of sources have been summarized previously.[1] These and other values are given in Table 6.1. The effect of temperature and of molecular weight on the specific gravity of low-molecular-weight PTHFs is given in Chapter 7.

6.3 MOLECULAR WEIGHT

PTHF of any desired molecular weight can be prepared. Methods of synthesis and their effect on the properties of the resulting polymer have been described repeatedly throughout this monograph (see Chapters 3–5 and 7). Oligomers (dimers, trimers) of very low molecular weight are liquids. Polymers with degrees of polymerization (DP) of about 10 are already sticky, viscous oils with a marked tendency to crystallize slowly at room temperature. At intermediate DPs, the polymers first become waxes that flow easily

TABLE 6.1
Density of PTHF

Type	T, °C	Density, g/ml	Ref.
Amorphous	20	0.982	1
Amorphous	25	0.990	1
	25	0.979[a]	2
In solution	20	1.02	1
Crystalline (x ray)	—	1.16–1.18	1, 3
		1.11	4
		1.09[b]	2
Crystalline (flotation)	—	1.07–1.08	1
Crystalline (weighing bottle)	—	1.06	5
	25	1.02	5
	25	1.025	5

[a]From specific volumes, V_s, determined in a density-gradient tube. The coefficient of expansion for the amorphous polymer, dV_s/dT, as a function of temperature, T, was 7.3×10^{-4} cm^3/(g °C).

[b]The temperature coefficient for the crystalline polymer was assumed to be 2×10^{-4} cm^3/(g °C).

on melting and then, with further increase in molecular weight, become tough, rubbery materials that retain their shape on melting. Such amorphous PTHFs have excellent green strength and good tack, but they lose much of this as they slowly crystallize at room temperature. When the molecular weight is in the millions, the crystalline polymers are tough and almost intractable.

Low-molecular-weight PTHF glycols and diisocyanates suitable for the preparation of polyurethanes or polyesters can be purchased from a number of suppliers. Some of these are given in Chapter 7. PTHF standards suitable for the calibration of GPC columns or for crystallization studies near room temperature can be obtained from Polymer Laboratories Ltd.[6] These PTHFs have the structure $CH_3CH_2\text{–}[O(CH_2)_4\text{–}]_nOCH_3$, have nominal molecular weights ranging from 1,000 to 300,000 and have molecular-weight dispersities of 1.10 to 1.20.

6.4 CRYSTALLIZATION

6.4.1 Morphology

PTHF is a partially crystalline polymer. It forms spherulites and lamellae and can be drawn into fibers. According to Fielding-Russell and Wetton,[7] the crystallites grow radially as spiralled lamellae from heterogeneous nucleation centers. Ringed spherulites have been observed after crystallization of a PTHF sample with $\bar{M}_v = 540{,}000$.[7]

The degree of crystallinity reported varies from 24% to 80%,[2,7–9] depending on the method of preparation of the polymer, its molecular weight and the temperature of crystallization. Lower-molecular-weight polymers generally have higher degrees of crystallinity. Higher crystallization temperatures give more perfect crystals and polymers with higher melting temperatures but lower degrees of crystallinity.

Fielding-Russell and Wetton[7] have obtained additional information on the crystallization process of PTHF from dielectric permittivity. They analyzed their data according to the Onsager relationship between permanent isolated dipole moments (μ), the dielectric permittivity (ϵ) and the refractive index (n) of condensed dipolar small molecules. A value of 2.55 based on the ϵ value at 1.4 Mc/s at −197°C was used for the n^2 term that appears in the equation. This suggests that PTHF itself has a refractive index of 1.5969, which is much higher than the value of 1.48 at 20°C given in Table 6.6 below.

Fielding-Russell and Wetton[7] found that the percentage of crystallinity obtained from their dielectric measurements correlated better with percentages obtained by specific-volume measurements of the primary crystallization process than it did with percentages obtained from x-ray diffraction studies. They suggest that the secondary crystallization process at the crystallization temperatures studied, 18° and 23.5°C, and at a single frequency of 1.4 Mc/s does not involve crystallization of the residual amorphous phase. Instead, an improvement in the order of the chains within the crystalline phase is proposed.

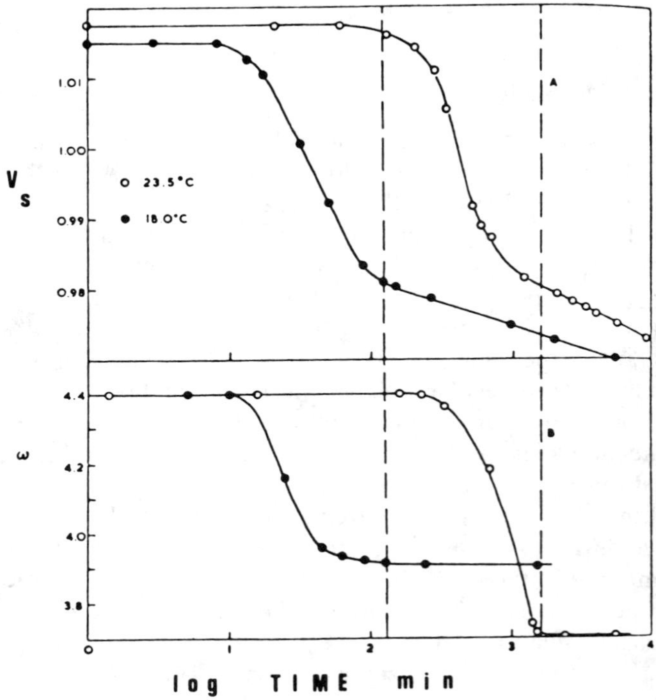

FIGURE 6.1 Crystallization of PTHF as a function of time at 18°C and 23.5°C followed by specific volume (A) and dielectric permittivity (B).[7]

It is beyond the scope of this chapter to discuss details of the analysis by dielectric permittivity, but some readers may be interested in the comparison shown in Figure 6.1 of results obtained by specific-volume measurements and by dielectric permittivity during crystallization of PTHF.

6.4.2 Structure of the molecule

The earliest report of the structure of the PTHF molecule by Bunn and Holmes[10] suggests a planar zigzag conformation. Subsequent IR and Raman spectroscopy studies[3,11] have confirmed this structure. Raman and polarized IR spectra of

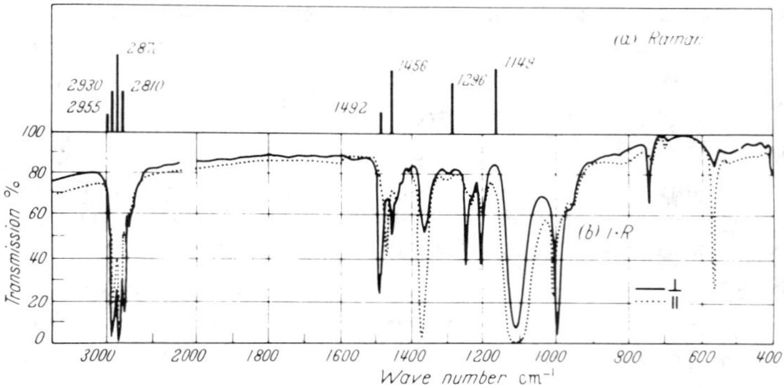

FIGURE 6.2 Raman spectrum (a) and polarized IR spectrum (b) of oriented PTHF. The solid and broken lines in (b) represent the spectrum measured with the electric vector perpendicular and parallel to the orientation direction, respectively.[11]

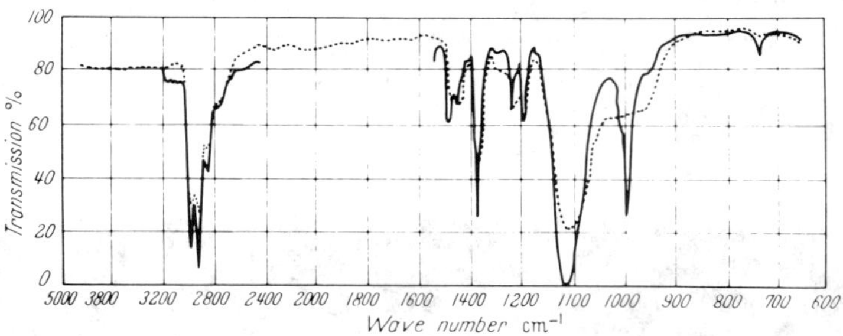

FIGURE 6.3 IR spectrum of PTHF, crystalline at 25° (continuous curve) and as a melt at 50°C (dashed curve).[12]

oriented PTHF are shown in Figure 6.2. Ordinary IR spectra of crystalline and amorphous PTHF are shown in Figure 6.3 for comparison.

6.4.3 Crystal structure of PTHF

Published knowledge about the crystal structure of PTHF is due principally to the x-ray studies of three groups.[3,11,13] Photographs of the x-ray diffraction patterns of unoriented PTHF typically display two strong rings whose diffraction peaks correspond to d-spacings of about 4.5 and 3.7 Å, respectively.[14] Photographs used for the structure analysis had as many as 17 diffraction peaks, although only two of them were strong.[3]

Vainshtein *et al.*[3] studied unoriented PTHF while Imada *et al.*[11] and Cesari *et al.*[13] studied oriented PTHF. An x-ray photograph of a highly doubly oriented PTHF and some assignments for the indices of the reflections are shown in Figure 6.4. The conclusions reached by the three different groups are compared in Table 6.2. Vainshtein *et al.* suggest a rhombic structure with the oxygen atoms all in register for unoriented PTHF. Their model is shown in Figure 6.5. Imada *et al.* suggest a monoclinic structure for oriented PTHF with two essentially planar zigzag chains running through the center and corners of the unit cell parallel to the *c*-axis. The zigzag planes of the molecular chains are parallel to the bc plane in the latter model. As shown in Figures 6.6 and 6.7, the oxygen atoms in this model are not in register, so that the electronegative oxygen atoms of each chain are located as far as possible from each other.

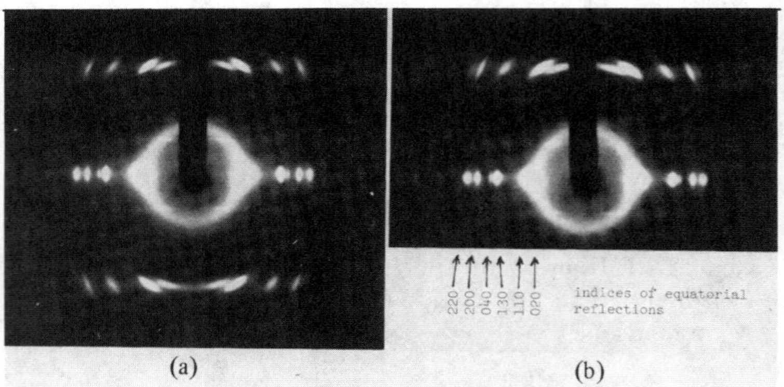

FIGURE 6.4 X-ray fiber photographs of PTHF taken at room temperature and at $-100°C$ (b) with monochromatized Cu K_α radiation. Also indicated are the 220, 200, 040, 130, 110 and 020 indices of equatorial reflections.[15]

TABLE 6.2
Crystallographic Data for PTHF from X-ray Studies

Orientation	Unit cell	a (Å)	b (Å)	C (Å)	β(°)	Space symmetry group	Ref.
Oriented	Monoclinic C2/c	5.61 ± 0.04	8.92 ± 0.05	12.25 ± 0.1 (fiber axis)	134.5	D_{2h}	13
Oriented	Monoclinic C2/c-C^6_{2h}	5.48	8.73	12.07 (fiber axis)	134.2	D_{2h}	11
Oriented	Monoclinic C2/c-C^6_{2n}	5.59	8.90	12.07 (fiber axis)	134.2	D_{2h}	15
Unoriented	Rhombic	12.25 ± 0.1	8.75 ± 0.1	7.22 ± 0.05	—	$P_{2_1 2_1 2_1}$	3

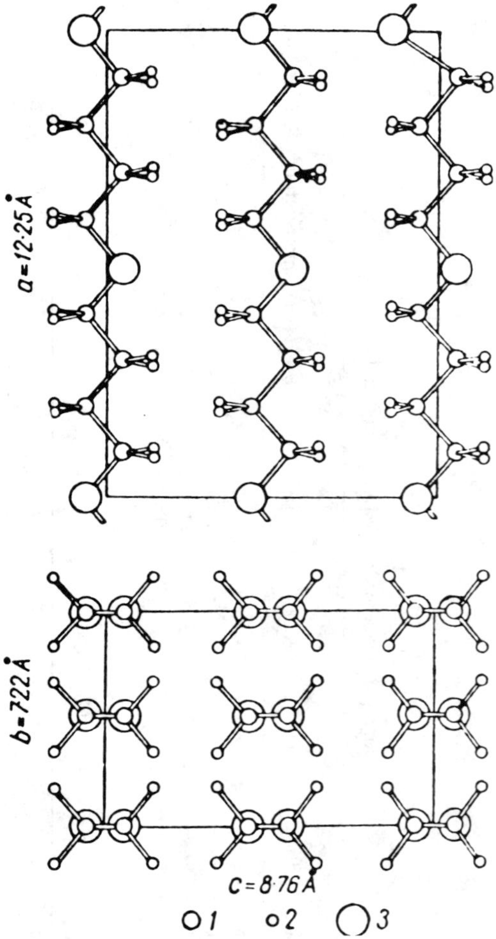

FIGURE 6.5 Model of unoriented PTHF specimen with space group $P_{2_12_12_1}$. In the figure 1 denotes carbon, 2 denotes hydrogen and 3 denotes oxygen.[2]

If all authors are correct in their interpretations, clearly the plane of the oxygen atoms tilts and the crystal structure changes on orientation. Further work on the crystal structure of PTHF is indicated.

FIGURE 6.6 The arrangement of two neighboring PTHF molecules in an oriented PTHF sample, viewed perpendicular to the bc plane.[11]

Another unexplained observation about the crystal structure of PTHF is that the breadths of equatorial reflections in the x-ray fiber diagrams at room temperature are not always the same.[15] That is, the hk0 reflection is relatively broad compared to the h00 and 0k0 reflections. Furthermore, x-ray diffraction patterns of PTHF taken at various temperatures from −190°C to room temperature also show that the breadth of the hk0 reflections increase remarkably at low temperatures.[15]

6.4.4 Melting temperature of PTHF

The melting temperature most often reported for PTHF is about 43°C,[2,8,13,16] although values as high as 58–60°C are also in the literature.[17]

Effect of chain ends The melting temperature of PTHF is clearly a function of molecular weight. This is evident from the

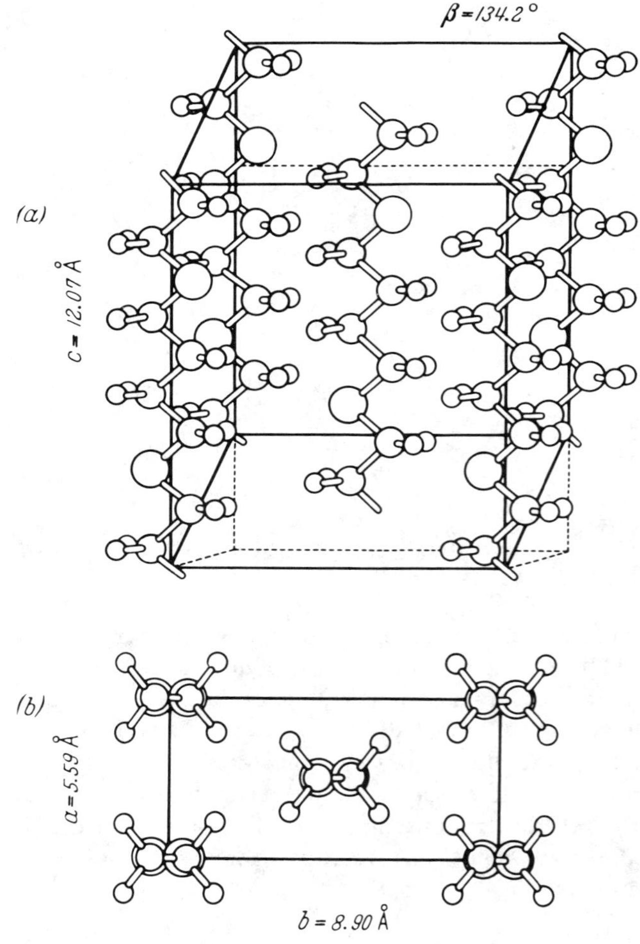

FIGURE 6.7 (a) The arrangement of polymer chains in the unit cell of oriented PTHF viewed roughly perpendicular to the fiber axis. (b) The packing of the molecules viewed down the fiber axis.[11]

data plotted in Figure 6.8,[18] where melting temperatures are plotted for polymers having molecular weights ranging from 750 to 12,800. Extrapolation of these data suggests an equilibrium melting temperature, T_m^0, of (43 ± 3)°C. This treatment

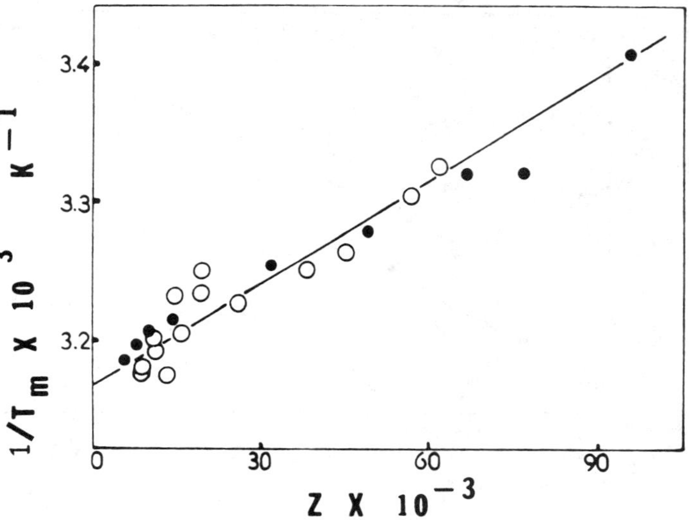

FIGURE 6.8 Reciprocal of melting temperature versus degree of polymerization, $Z = \bar{M}_n/72$, for hydroxy-terminated PTHF, unfractionated (filled circles) and fractionated (open circles).[18]

based on chain ends as diluents is known to give unreliable values for the heats of fusion but the extrapolated value of T_m^0 should be more reliable. In contrast, Trick and Ryan[2] have used two different methods to determine the melting temperature of PTHF using a polymer with $\bar{M}_n$ of ~130,000 in their studies; their estimate of T_m^0 is about 51°C. In making their extrapolation they ignored what appeared to be an anomalously high melting temperature for a high-molecular-weight polymer crystallized at about 25°C (cf. Figure 6.9[2]).

Even this T_m^0 determined by Trick and Ryan is probably too low for the T_m^0 of PTHF. Sorenson and Campbell have reported a value of 55°C for the melting temperature of a high-molecular-weight polymer prepared with PF_5 as initiator.[19] Moreover, melting temperatures above 50°C have routinely been observed in our laboratories on the first heating of measurements using differential scanning calorimetry.[20] Lower

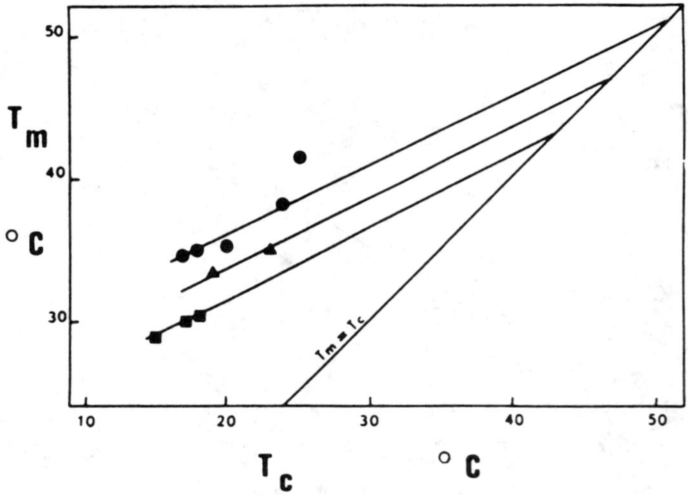

FIGURE 6.9 Variation of melting temperature, T_m, with temperature of crystallization, T_c. Filled triangles correspond to polymer A, filled circles to polymer B, and filled squares to polymer C.[2]

melting temperatures are observed on the second and third cycles. It appears that either molecular weight and temperature of crystallization have a remarkable effect on the melting temperature, or that more than one crystal structure of PTHF is possible.

Effect of solvent Crystallization of PTHF from toluene solution gave anomalous results.[2] The most dilute solutions (0.09%) produced a two-phase system containing crystals whose disolution temperature, T_s, was markedly dependent on the temperature of crystallization, T_c. Crystallization from less dilute solutions (1% and 3%) led to a single, gel-like phase. Formation of gel was most pronounced at higher concentrations and molecular weights of PTHF and at higher supercoolings (lower T_c). On melting, the gel structure disappeared at about −25°C, but the T_s values were not a simple function of either concentration of PTHF or T_c.

There is a clear need to critically reexamine the structure of PTHF obtained by x-ray and the melting temperature results, with careful attention to the effects of molecular weight, crystallization temperature, solvent and degree of orientation.

Effect of copolymeric units As usual, the effect of copolymeric units is to lower the melting temperature. For example, copolymerization with 3,3-*bis*(chloromethyl)oxetane (BCMO) or 2-methyltetrahydrofuran gives amorphous products.[12,21] Saegusa *et al.* found that the infrared spectrum of the copolymer with BCMO was similar to a combination of those of the homopolymers, except that the absorption at 1000 cm^{-1} characteristic of crystalline PTHF was absent.[12] In another study,[22] random copolymers of THF and propylene oxide (PO) were prepared using BF_3·THF at concentrations below 0.15%. The copolymers showed decreasing melting temperatures as the content of PO increased. No crystallization was obtained, when the mole percent of units from PO was greater than 40%.[22] At higher contents of BF_3·THF, the crystallinity of the resulting copolymers increased sharply as a result of uneven distribution of units from PO in the copolymer.[22] Additional information about the melting temperature of PTHF in block and graft copolymers is in Chapters 7 and 8. This chapter is primarily concerned with properties of the homopolymer and of random copolymers containing PTHF.

Effect of pressure Kijima *et al.*[23] have studied the effect of pressure on the melting temperature of a PTHF sample prepared at 1–2°C using $BF_3 \cdot Et_2O$ and having $\bar{M}_v = 83,000$. Melting temperatures at atmospheric pressure were determined by differential thermal analysis. A high-pressure dilatometer was used to determine melting temperatures under pressure. Slow cooling of a PTHF sample to 0°C under one atmosphere led to a polymer with a melting temperature of 38°C and a pressure coefficient of 0.017°C $(kg/cm^2)^{-1}$. Melting at atmospheric pressure of a sample crystallized for 8.5 hours at 69.0°C under 4800 kg cm^{-2} led to a DTA thermogram with *three* melting peaks, one each at 34, 39 and 44°C. Wide-angle x-ray scattering patterns of the two polymers were the same. The small-angle x-ray scattering (SAXS) pattern of the sample crystallized under pressure showed a fairly sharp ring corresponding to a long period of about 200 Å, while that of the sample crystallized at atmospheric pressure showed a diffuse ring corresponding to a nearly identical long period. Kijima *et al.*[23] concluded that the different melting temperatures could be ascribed to crystals with different lamellar thicknesses, and that perhaps the highest melting temperature could be attributed to extended or nearly extended chain lamellae, which are incapable of detection with SAXS. Such an interpretation seems improbable, since PTHF with even higher melting temperatures can be obtained from PTHFs crystallized at atmospheric pressure.

The effect of pressure on the melting temperature of PTHF, as observed in the high-pressure dilatometer, is shown in Figure 6.10. As usual, at high pressures the increase in melting temperature is substantial.

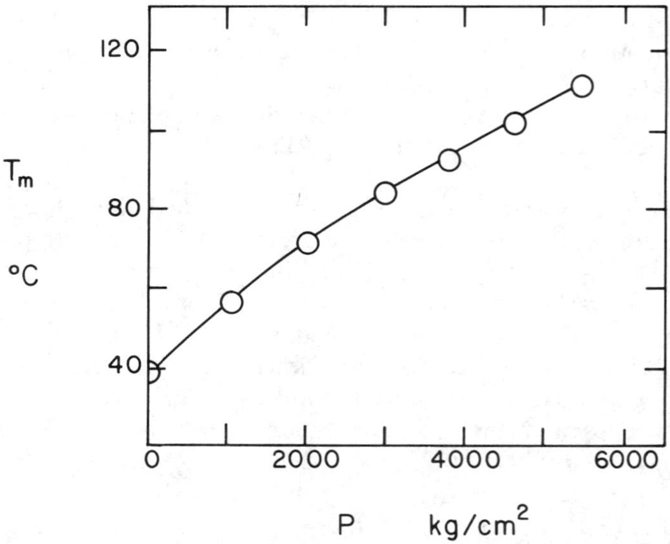

FIGURE 6.10 Effect of pressure on the melting temperature of PTHF.[23]

Concluding Comments There is ample evidence to suggest the existence of two or perhaps three crystal structures for PTHF. One structure could be responsible for the formation of gels such as those now confirmed for isotactic polystyrene.[24] By analogy with this work, such a crystal would be of the fringed-micelle type. Another might form upon crystallization from the bulk, especially of low-molecular-weight polymers, at critical temperatures near room temperature or below, and still a third may form from high-molecular-weight polymers isolated by evaporation of THF and/or other solvents crystallized above, say, 30°C. There exists a strong case that the T_m values of ~43°C reported so frequently are real, and are not attributable either to impurities in the PTHF or to the use of PTHF of exceptionally low molecular weight. Similarly, the higher values of 50°C and above are also real. Possible reasons for the discrepancies among the results of various workers are under investigation in our laboratory.

6.4.5 Kinetics of crystallization

Some information about rates of crystallization obtained by specific-volume and dielectric-permittivity measurements has already been given in Section 6.4.1.[7] Other studies have been made by dilatometry[2] and by infrared spectroscopy.[25] The crystallization isotherms have the typical S-shaped appearance familiar to crystallization kineticists.

Trick and Ryan[2] used dilatometry for their studies and examined the rates of crystallization of PTHFs of two different molecular weights and of one copolymer. They compared the results, again as usual, in terms of the time, $t_{1/2}$, required to reach half the final extent of crystallinity (measured as half the final extent of volume contraction). Some of their data are reproduced in Table 6.3. The lower-molecular-weight polymer crystallized faster and attained a higher degree of crystallinity. The introduction of copolymeric units decreased the rate of crystallization and the degree of crystallinity. Attempts to fit the Avrami equation to the experimental data in the early stages of crystallization led to a tentative value of $N = 3$ for

TABLE 6.3
Half Times of Crystallization[2]

T_c, °C	Polymer A[b]	$t_{1/2}$, min[a] Polymer B[c]	Polymer C[d]
15.0	—	—	55
17.0	—	—	175
18.0	—	34	318
19.0	42	66	630
20.0	—	99	—
21.0	105	204	—
23.0	340	700	—
24.0	—	1,200	—
25.2	1,740	—	—
25.8	—	5,160	—
27.0	—	14,400	—

[a]Samples were melted for 30 min at 60°C before crystallizing. Studies showed that melting at temperatures ranging from 50° to 100°C did not affect the results.

[b]Polymer A had $\bar{M}_n = 6{,}760$ by end-group analysis.

[c]Polymer B had $\bar{M}_n = 130{,}000$ by osmometry.

[d]Polymer C was prepared from a glycol with $\bar{M}_n = 6{,}630$ by first phosgenating and then chain extending.

the exponent in the equation. This suggests the growth of instantaneously nucleated spherulites.[4]

Rushchupkin *et al.*[25] were interested in comparing the rates of crystallization of PTHF with those of polyethylene oxide (PEO). Their infrared studies indicated that at a given supercooling the rate of crystallization of PEO is much higher than that of PTHF. They indicated that data from IR regarding structure of the melts and rates of segmental motion measured by dielectric relaxation show that the difference stems from higher order in the PEO melt than in the PTHF melt. As a result, the rate of nucleation and hence of the overall rate of crystallization is higher for PEO. According to these workers the rates of growth of the crystallites are similar.[25]

6.4.6 Thermodynamics of crystallization

Huglin *et al.*[18] quote values of 21.4 J deg^{-1} mole^{-1} and 6.8 kJ mole^{-1} for ΔS_u and ΔH_u of fusion per mole of repeat unit of PTHF, respectively. The ΔS_u value is based on a T_m^0 of (43 ± 3)°C and the relationship $T_m^0 = \Delta H_u/\Delta S_u$.[26] Other literature values of ΔS_u and ΔH_u based on calorimetry were 34.8 J deg^{-1} mole^{-1} and 11.0 kJ mole^{-1}, respectively,[27] and those based on measurements in a special type of differential-thermal-analysis apparatus were 39.4 J deg^{-1} mole^{-1} and 12.4 kJ mole^{-1}, respectively.[28]

6.4.7 Crystallization during oligomerization

Stejny[29] has studied the polymerization of THF initiated by $Et_3O^{\oplus}SbCl_6^{\ominus}$ at temperatures below 0°C. He found that, during polymerization of THF in bulk at −25°C and below, PTHF crystallizes from a solution in its own monomer. The crystallization causes a slight acceleration in the polymerization rate and begins at a time when the polymer molecular weight and concentration are 9,000 and 15%, respectively. The corresponding values for polymerization at −45°C are 3,900 and 2%. Separate crystallization experiments from solution were carried out with PTHFs having $\bar{M}_n$ similar to that during polymerization. Two different solvents and crystallization temperatures were used: THF with a T_c of −45°C; and a mixed solvent consisting of isopropanol and 2-butanol at 0°C. The precipitated crystals were collected into a mat at the crystallization temperature and then dried. X-ray studies at room temperature indicated that the crystals were lamellae with a thickness of 110–22 Å. From this result, Stejny concluded that lamellae with folded chains form during crystallization.[29]

6.5 GLASS-TRANSITION TEMPERATURE

There is general agreement that PTHF forms a glass at about −86°C.[2,8,19,30–32] Low-molecular-weight glycols may do so at a somewhat higher temperature.[33,23] Comonomers at low concentration have little effect on T_g [35] but, as noted in Section 6.4.4, can markedly decrease the melting temperature.[35] The effect of block- and graft-copolymer formation on T_g is discussed in Chapter 8.

The effect on T_g of adding $ZnCl_2$ to a solution of PTHF glycol with $\bar{M}_n = 2010$ and then evaporating the solvent is shown in Table 6.4. The T_g is slightly increased, but the increase is much less than that observed with similar concentrations of $ZnCl_2$ in PPO.[33,34] The marked depression of the melting temperature is approximately in accord with the theoretical depression of T_m predicted by Flory for a randomly dispersed diluent in a polymer.[26] Furthermore, thermal data also indicate that crystallinity decreases until about 15 mole% $ZnCl_2$, when the PTHF becomes completely amorphous.[33,34] Wetton and coworkers[33,34] explain the results for PPO in terms of a chelate-ring model involving coordination of two adjacent ether oxygen atoms. A similar structure for PTHF is entropically unfavorable and only intermolecular crosslinking occurs.

TABLE 6.4
Effect of $ZnCl_2$ on the T_g and T_m of PTHF[33,34]

$ZnCl_2$, mol%	T_g, °C	T_m, °C
0	−76	24
1.76	−76	24
3.91	−76	23
8.01	−70	20
14.7	−65	14
28.7	−63	—[a]

[a]The polymer did not crystallize.

6.6 DEGRADATION AND STABILIZATION

On the one hand, the saturated backbone of PTHF makes it less susceptible to certain kinds of degradation than hydrocarbons containing unsaturation along the backbone. On the other hand, the presence of oxygen atoms regularly spaced along the backbone makes PTHF particularly susceptible to oxidative degradation. The degradation of PTHF has been studied. It clearly occurs and stabilizers are known. In addition, when reactive end groups are present, appropriate precautions must be taken to prevent degradative or cross-linking reactions. Reactions that can occur when hydroxyl end groups are present have already been described in Chapter 5 and will not be repeated here. Reactions that can occur with isocyanate end groups are described in Chapter 7. In this chapter only reactions that can occur along the backbone are described.

6.6.1 Degradation of the bulk polymer

Davis and Golden[36] have studied the degradation of PTHF in bulk at various temperatures. The polymers studied were prepared using $PF_5 \cdot THF$ either in an open flask (polymer A) or in vacuum with exposure to air during the workup (polymer V). The intrinsic viscosity of polymer A, heated at constant temperature up to 150°C in a sealed system, fell rapidly to a constant value. Polymer V behaved similarly, but the decrease was considerably smaller. When heated in air at constant temperature, the viscosity of both polymers decreased continuously until the polymer was completely degraded. Temperatures well above 150°C were needed to degrade the polymers completely in vacuum.

The rate of weight loss of polymer A was determined in vacuum and in air. The activation energy for degradation in vacuum calculated from the data was 45 kcal mole^{-1}. This value is comparable with those of PPO (46 kcal mole^{-1}), but low compared to hydrocarbon polymers for which the activation energy for degradation is 60–70 kcal mole^{-1}. The corresponding value for degradation of PTHF in air was 29 kcal mole^{-1}.

Neiman *et al.*[37] have studied the kinetics of thermooxidative degradation at 90–120°C of a PTHF having a molecular weight of 1130. Hydroperoxides were formed at all stages of the degradation. Kinetic curves of the oxidation showed an autoacceleration according to the formula

$$-\Delta p = A\exp(\Phi t),$$

where Δp is the pressure change in the system; A is a constant; Φ is the coefficient of acceleration; and t is time. The effective energy of activation was 15 kcal mole^{-1}. The oxidation could be inhibited by adding amines and pyrocatechol.

In a study of the stabilization of copolymers of propylene oxide and THF against oxidative degradation, Okunev[38] examined the effect of 25 antioxidants including mono and bis-phenols, amines, quinones, H_3PO_4 and others. The most effective antioxidants at 120–50°C and an oxygen pressure of 53 kPa were phenothiazine, hydroquinone, phenyl-2-naphthylamine and 1-naphthyl pyrocatechol phosphate. The presence of hydroperoxide groups in the polymers decreased the effectiveness of the antioxidants.

6.6.2 Degradation in solution

Davis and Golden[36] found that the intrinsic viscosity of a benzene solution heated in air dropped steadily in the absence of antioxidant. In the presence of an antioxidant such as 0.5% 2,6-di-*t*-butyl-4-methylphenol, the intrinsic viscosity fell initially, but thereafter became constant. As shown in Figure 6.11, this initial drop was greater for polymer A than for polymer V. Degradation was accompanied by the appearance of a strong aldehydic carbonyl band in the infrared. Davis and Golden[36] postulated that the weak links are peroxide groups formed during contact of the polymer with air. The mechanism they proposed for degradation was:

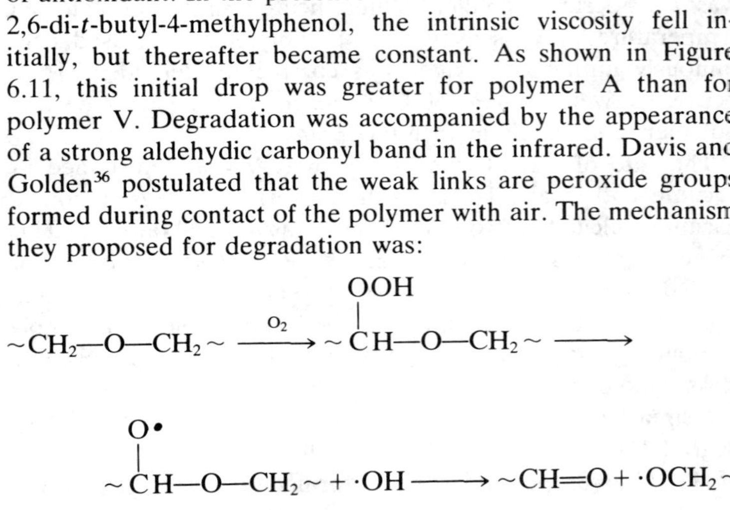

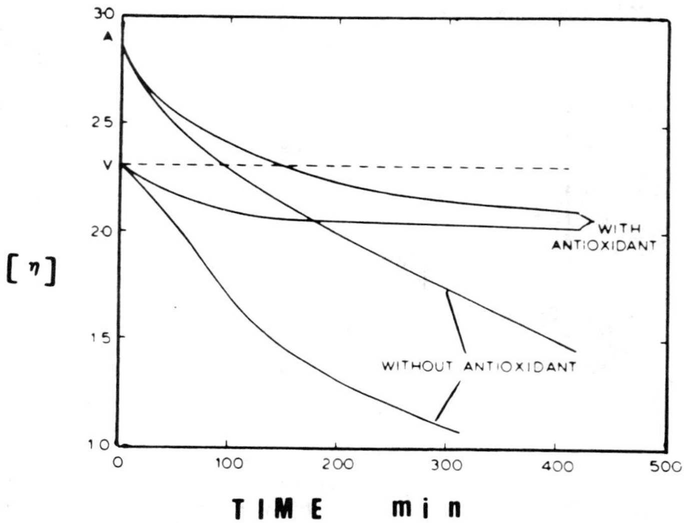

FIGURE 6.11 Effect of time of heating on intrinsic viscosity, [η], dl/g, of xylene solution of PTHF prepared in air (A) or in vacuum (V). Polymer concentrations were 0.3% When used, the antioxidant concentration was 0.5%.[36]

At higher temperatures in the absence of air the degradation mechanism was supposed to be random scission occurring primarily at the C–O bond:

$$RCH_2OCH_2R \longrightarrow RCH_2O\cdot + \cdot CH_2R$$

6.6.3 Degradation and crosslinking due to ionizing radiation

PTHF is not stable to ionizing radiation. Analysis of the solubility behavior of PTHF after exposure to high-energy electrons in vacuum[39] indicated a ratio of main-chain scissions to crosslinks of 0.37. Hydrogen and traces of other gases were evolved. The relationship between intrinsic viscosity and radiation dose is shown in Figure 6.12. The behavior of PTHF toward ionizing radiation is similar to that of PPO, which also undergoes both chain scission and crosslinking. It is different

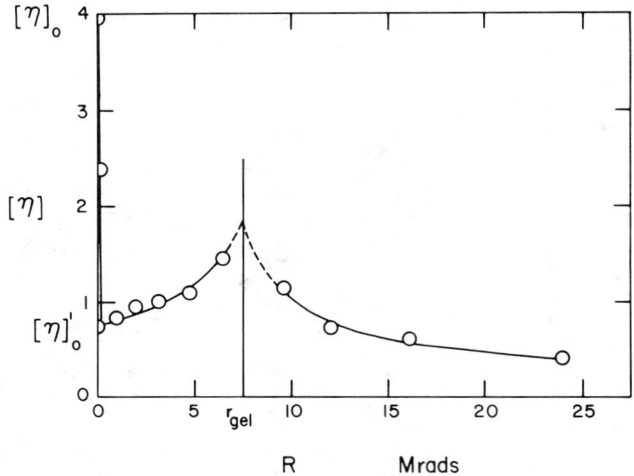

FIGURE 6.12 Relationship between intrinsic viscosity, $[\eta]$, and radiation dose R, where $[\eta]_0$ denotes the intrinsic viscosity, $[\eta]_0'$ denotes the effective intrinsic viscosity after the destruction of weak links and r_{gel} denotes the radiation dose at gel point. The ratio of chain scissions to crosslinks was 0.37. All intrinsic viscosities are in dl/g.[39]

from that of polymethylene oxide, which undergoes only chain scission, and of polyethylene, which crosslinks.

Thomassin and Marchal[40–43] have published a series of papers describing the primary reactions occurring during radiation induced autooxidation of PTHF in chloroform solution at 25°C. Two primary reactions were proven. Both involve preliminary formation of peroxy radicals after abstraction of a hydrogen from the polymer.

The first is a decomposition of α-alkoxyalkylperoxy radicals by a unimolecular mechanism to give ethylene and other macromolecular fragments:

$$PO_2\cdot \longrightarrow P'{=}O + H_2C{=}CH_2 + P'O\cdot$$

(Except for the formation of ethylene this mechanism is similar to the one proposed by Davis and Golden for degradation in solution.) Some of the ethylene is oxidized to form ethylene oxide, which partly inhibits the chain scission.

The second reaction is a nonterminating bimolecular reaction of the same initial radical with other peroxy radicals in the medium. This reaction produces secondary α-alkoxyalkyl radicals which decompose by β-scission:

$$PO_2\cdot + RO_2\cdot \longrightarrow PO\cdot + RO\cdot + O_2$$

$$PO_2\cdot + PO_2\cdot \longrightarrow 2PO\cdot + O_2$$

$$PO\cdot \longrightarrow P\cdot'' + P''{=}O$$

The probabilities of hydrogen-atom abstraction, monomolecular decomposition and nonterminating bimolecular reaction were 0.2–0.4, 1 and 7–10, respectively.

6.7 SPECTRA

The application of Raman and infrared spectra for the determination of the structure of the PTHF molecule has already been discussed in Section 6.4.2.

The occurrence of low-frequency Raman-active longitudinal acoustic modes in PTHF has been demonstrated.[44,45] The presence of these modes offers the possibility of using Raman spectroscopy to identify and estimate the thickness of PTHF lamellae[44] and to study the effect of hydrogen bonding on the lamellar surface.[45]

NMR spectra of PTHF can be very informative. The value of ^1H, ^{13}C, ^{19}F and ^{31}P NMR in kinetic studies and for end-group analysis was described in Chapters 3 and 4, where the appropriate chemical shifts and references were described.

NMR spectra of drawn and rolled sheets of PTHF have been taken from $-196°$ to 37°C and at various arrangements in the magnetic field.[46] The NMR measurements confirmed that in drawn and rolled sheets the crystallites orient uniaxially and doubly, respectively. A reduction of the second moment in the crystalline regions with increasing temperature was observed, and attributed to molecular motions within the crystal.[46]

NMR is also useful for determination of copolymer composition. One example is the analysis of copolymer microstructure of random copolymers formed during the cationic copolymerization of THF with diketene, $CH_2{=}\underline{CO\overset{\overset{O}{\|}}{C}C}H_2$.[47] The fraction of addition to the double bond was 0.37–0.41.

6.8 OTHER THERMODYNAMIC PROPERTIES

Having already noted some thermodynamic properties related to the polymerization of THF (Chapter 4) and the crystallization of PTHF (this chapter), in this section, we consider other thermodynamic properties.

Tsujita *et al.*[48,49] have measured a variety of thermodynamic properties for PTHF. Included among them are the internal pressure, P_i, the thermal expansion coefficient, $\alpha = (1/V)(\partial V/\partial T)_P$, the compressibility, $\beta = -(1/V)(\partial V/\partial P)_T$, and P–V–T relations from $-40°$ to 170°C.[48] The properties did not vary linearly with temperature and had discontinuities in the region of melting. Complete disclosure of their results would be prohibitively long, but a few results can be stated. They found that α was in the range $4–7 \times 10^{-4}$ deg^{-1}, β was in the range $4–10 \times 10^{-5}$ (kg/cm^2)$^{-1}$ and $P_i \approx 75$ cal/cm^3 at -35°C. The change in internal pressure on melting was positive and approximately equal to 67 cal/cm^3. The latter disagrees with earlier results of Sims, who found a negative change in P_i when the polymer melted.[50]

Yoshida *et al.*[28] have measured heat capacities for annealed and rapidly cooled PTHF samples in the temperature region from $-262°$ to 57°C using an adiabatic calorimeter. They found that ΔC_p at T_g was 19.4 and 15.8 J deg^{-1} mole^{-1} for rapidly cooled and annealed samples, respectively. Other determinations of thermodynamic parameters of PTHF have been made by Lebedev and Lityagov[51] and by Clegg *et al.*[52]

6.9 MECHANICAL PROPERTIES

6.9.1 Mechanical properties compared to thermoplastics

Burrows[53] compared a few of the mechanical properties of a PTHF produced using BF_3 with typical values for thermoplastics. Some of his data are given in Table 6.5. PTHF is much more extensible than most plastics.

TABLE 6.5
Representative Mechanical Properties of Selected Thermoplastics[53]

Polymer	Tensile strength, MPa	Modulus of elasticity, MPa	Elongation, %
PTHF	29.0	97	820
Poly(ethylene)—low density	13.8	170	500
Poly(ethylene)—linear	24.1	790	50
Poly(tetrafluoroethylene)	27.6	4000	200
Nylon 66	69.0	3100	45
Poly(BCMO)	44.8	860	100

6.9.2 Mechanical properties compared to elastomers

A comparison of the properties of a high-molecular-weight PTHF with those of familiar elastomers is shown in Table 6.6. Amorphous PTHF has good tack and good green strength and can easily be processed as a rubber. Crystalline PTHF of moderately high molecular weight behaves more like a thermoplastic.

6.9.3 Dynamic mechanical measurements

Wetton and Allen[57] have measured the storage and loss components of the rigidity modulus, G' and G'', respectively, for PTHF in the frequency range 0.01 to 1000 cycles s^{-1} at temperatures from −150° to +100°C. They used a specially designed apparatus. The results that are obtained at a frequency of 100 cycles s^{-1} are shown in Figure 6.13.[1] PTHF has two definite

TABLE 6.6
Properties of Some Common Rubber Gum Stocks Compared to PTHF[54–56]

Rubber	Shore A hardness	Specific gravity (25°C)	Refractive index, n_D^{20}	Dielectric constant at 1 kHz/sec and 25°C	T_g, °C	Melting temp., °C	Tensile strength, MPa	Elongation, %
PTHF	95	0.99–1.18	1.48	5.0	−86	54	27.6–41.4[a]	300–600[a]
Hevea	30–100	0.92	1.52	2.7	−74	6(32)	>20.7	780
SBR	40–100	0.94	1.53	2.9	−64	−35	<6.9	270
Butadiene–acrylonitrile	20–100	1.02	1.52	13.0	−23	—	>4.1	—
Neoprene	40–95	1.25	1.56	9.0	−50	32	>20.7	875
Butyl	40–75	0.91	1.51	2.4	−75	—	>10.3	675
Silicone	20–90	0.97	1.40	3.1	−123	−52	0	—

[a]Exact value depends on molecular weight.

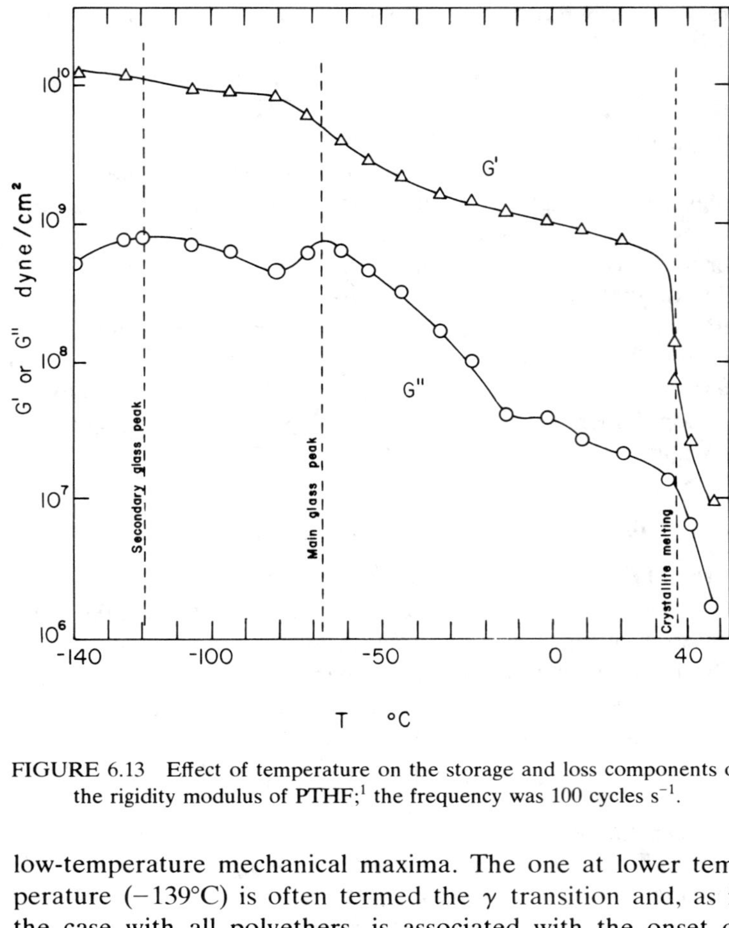

FIGURE 6.13 Effect of temperature on the storage and loss components of the rigidity modulus of PTHF;[1] the frequency was 100 cycles s^{-1}.

low-temperature mechanical maxima. The one at lower temperature (−139°C) is often termed the γ transition and, as is the case with all polyethers, is associated with the onset of micro-Brownian motion of the main chain. The other at −79°C is at the glass-transition temperature and is termed the β transition.[57] The storage rigidity-modulus increment for the γ transition is 3.5×10^9 dyne cm^{-2} and that for the β-transition is 7.2×10^9 dyne cm^{-2}. In the region of the melting temperature, about 38.5°C, a dramatic drop in modulus appears.

Perena and Fatuo have made similar measurements using a Rheovibron dynamic viscoelastometer.[58] Their measurements are in the range from $-160°$ to $+20°C$. The γ relaxation ranged from $-120°C$ at 110 Hz to $-138°C$ at 3.5 Hz and had an activation energy of 8 ± 1 kcal mole^{-1}. The β-relaxation ranged from $-69°C$ at 110 Hz to $-62°C$ at 3.5 Hz, with an activation energy of 94 ± 30 kcal mole^{-1}.

Viscoelasticity and other mechanical properties have been measured for high-molecular-weight copolymers of THF and BCMO[59] or 3-ethyl-3-chloromethyloxetane (ECMO).[60] Properties varied systematically with composition. Copolymers containing BCMO were more elastic but weaker when the sequences from THF alone were longer.[59] Copolymers containing ECMO did not crystallize when 2–30% of the copolymerized units were derived from the ECMO. The elasticity of the copolymer was substantially improved at lower degrees of crystallinity.[60]

6.9.4 Mechanical properties of the vulcanizates

Pure PTHF has no unsaturation and thus cannot be cured using conventional sulfur-cure recipes. Either peroxide or peroxide–sulfur cures have been used.[61–64]

Sulfur cures can be and have been used to cure copolymers of THF, when the comonomer is another cyclic oxide containing unsaturation in a side chain.[35,65–67] Many comonomers which have been used for this purpose were listed in Section 3.4.

PTHF has also been made UV-curable by copolymerizing with cyclic ethers such as 4-glycidyloxybenzophenone, glycidyl cinnamate or similar compounds.[68]

Some properties of cured PTHF are given in Table 6.7.[61,69] The properties are quite respectable, but not exceptional compared to vulcanizates from other, much less expensive, elastomers. The properties given are based on a PTHF having a molecular weight in excess of 3000. Lower-molecular-weight PTHF did not cure well with peroxide or peroxide–sulfur recipes. PTHF with a moderately high molecular weight in the tens or hundreds of thousands is best for curing by this method.

TABLE 6.7
Properties of Peroxide- or Peroxide–Sulfur Cured PTHF Stock and Comparison with Other Vulcanizates

Recipe[a]		300% modulus, MPa	Tensile strength, MPa	Elongation, %	Ref.
1. PTHF	100	1.6	25.5	650	61
Di-Cup 40C[b]	10				
Sulfur	1				
2. PTHF	100	7.4	38.3	560	61
Varox[c]	5				
Sulfur	1				
3. PTHF[d]	100	14.3	16.8	400	61
Carbon black (HAF)	50				
Dicumyl peroxide	4.8				
4. PTHF[d]	100	9.4	22.9	590	61
Carbon black (HAF)	50				
Dicumyl peroxide	4.8				
Sulfur	2.2				
5. PTHF[d]	100	11.2	36.9	740	61
Carbon black (MPC)	40				
Dicumyl peroxide	5.0				
Sulfur	1.4				
6. PTHF[d]	100	11.8	35.6	710	61
Carbon black (MPC)	40				
Varox[c]	4.5				
Sulfur	1.0				

Table 6.7 (*continued*)

Recipe[a]		300% Modulus, MPa	Tensile strength, MPa	Elongation, %	Ref.
7. Natural rubber[e]	100	2.1	29.6	730	69
Stearic acid	2				
Zinc oxide	5				
AGERITE STALITE	1				
Sulfur	2.75				
ALTAX	1				
TUADS	0.1				
8. Natural rubber[e]	100	22.6	23.0	310	69
Stearic acid	2				
Zinc oxide	5				
AGERITE STALITE	1				
Sulfur	2.75				
ALTAX	1				
TUADS	0.1				
Carbon black (HAF)	42				
9. SBR[f]	100	—	20.3	230	69
BONDOGEN	2				
Stearic acid	2				
Zinc oxide	5				
AGERITE STALITE	1				
Sulfur	2				
ALTAX	1.5				
CUMATE	0.1				
Carbon black (HAF)	42				

10. Neoprene W[g]	100	—	24.7	220	69
AGERITE STALITE	2				
Magnesia	4				
Stearic acid	0.5				
Zinc oxide	5				
THIATE E	0.5				
Carbon black (HAF)	28				

[a]Samples without carbon black were cured 60 min. Samples containing carbon black were cured 10 min; longer cure times led to poorer properties.

[b]40% dicumyl peroxide supported on calcium carbonate is a product of Hercules, Inc.

[c]Varox is the trade name of a commercial product containing 50% 2,5-ditertiarylbutyl-peroxy-2,5-dimethylhexane.

[d]The solvent was not specified but molecular weight was defined in terms of inherent viscosity. The PTHF in recipes 3 and 4 had $\ln \eta_r/c = 2.7$. The PTHF in recipes 5 and 6 had $\ln \eta_r/c = 3.0$. The time in which work was done suggests that most probably the solvent was benzene or toluene.

[e]Smoked sheet pressed 15 min at 142°C for optimum cure.

[f]Pressed 20 min at 142°C for optimum cure.

[g]Pressed 20 min at 152°C for optimum cure.

TABLE 6.8
Selected Physical Properties[35] of Cured Plasticized PTHF[a]

	Comonomer						
Property	4-vinyl cyclohexene oxide			4,5-epoxy-1-hexene		5,6-epoxy-1-hexene	
Iodine number[b]	10.5	15.6	24.7	10.5	16.3	10.5	16.3
Dioctyl azelate, phr	100	75	75	100	100	100	75
300% modulus, MPa	5.8	7.7	9.0	6.7	9.1	7.0	10.5
Tensile strength, MPa	16.6	15.6	17.4	13.7	14.1	16.7	19.0
Elongation, %	735	565	560	530	450	615	500
Dynamic modulus, MPa	4.3	6.7	9.5	4.6	6.3	4.1	6.9
Pico abrasion, 70°C	95	112	136	63	59	59	85
Ring flex—70°C, hours	166	40	18	65	10	70	20
Gehman freeze, °C	−90	−90	−90	−91	−91	−90	−93

[a]The plasticized PTHF was sulfur-cured using a compounding recipe including copolymer (100), dioctyl azelate (75–100), Irganox 1010 (2.6), N-330 Black (96–112), Goodrite 3146 (1), zinc oxide (4.5–5), stearic acid (4.5–5), tetramethylthiuram disulfide (1.5–2.5), dithiodimorpholine (1.5–2.5) and sulfur (2.5–3.5). The numbers in parentheses are in parts per hundred of rubber (phr).

[b]Iodine number is a property of the uncured rubber.

Vulcanizates have also been obtained by curing primary-amine-terminated PTHF of $\bar{M}_n = 10{,}000$ with diepoxide resins.[70] The vulcanizates had good tensile strength, elongation and abrasion resistance. They exhibit good adhesion to most metals, glass, fabrics and elastomers, and have good water resistance.

Dreyfuss and Macey[35] have explored the possibility of plasticizing high-molecular-weight PTHF as a means of reducing its tendency to crystallize. Their intent was to extend the useful temperature range of the high-molecular-weight elastomer. They found that dioctyl azelate and tricresyl phosphate were excellent plasticizers; but the plasticized homopolymer could not be cured. When copolymers containing some unsaturation were prepared, curing was possible and vulcanizates with good physical properties were obtained (see Table 6.8). However, as temperature is lowered, even the plasticized copolymers tended to stiffen, probably as a result of crystallizability of the PTHF. Once again, the products had no apparent advantage over vulcanizates prepared from conventional, less expensive elastomers.

6.10 MISCELLANEOUS APPLICATIONS

All the important commercial uses for PTHF employ low-molecular-weight (1000–3000) materials, generally in the form of diols. These materials are discussed in Chapter 7.

No significant market for unmodified high-molecular-weight PTHF is known. Some possible uses are known, however. Crystalline PTHF has been blended with *cis*-polybutadiene and used to make golf balls.[71] High-impact resins have been prepared from blends with poly(vinyl chloride) or chlorinated poly(vinyl chloride).[72] PTHF shows a certain resistance to the flow of salt and has been suggested as a possible membrane for desalination experiments.[73] It can be used to make composite membranes through which no nitrogen gas permeates at a pressure difference of 600 torr.[74,75]

REFERENCES

1. P. Dreyfuss and M. P. Dreyfuss, *Adv. Polym. Sci.*, **4**, 528 (1967).
2. G. S. Trick and J. M. Ryan, *J. Polym. Sci. C*, **18**, 93 (1967).
3. E. F. Vainshtein, M. Ya. Kushnerev, A. A. Popov and S. G. Entelis, *Vysokomol. Soyed.*, **A11**, 1606 (1969) (in Russian). Translated in *Polym. Sci. USSR*, **11**, 1820 (1969).
4. H. Tadokoro, in "Macromolecular Reviews", Vol. 1, A. Peterlin, M. Goodman, S. Okamura, B. H. Zimm and H. F. Mark (eds.), Interscience, New York, NY, p. 119 (1967).
5. M. B. Huglin and D. J. Pass, *J. Appl. Polym. Sci.*, **12**, 473 (1968).
6. "PLgel Columns, PL Polymer Standards", Polymer Laboratories, Ltd., Essex Road, Church Stretton, Salop SY6 6AX, UK (undated).
7. G. S. Fielding-Russell and R. E. Wetton, *Polym. Letters*, **5**, 761 (1967).
8. R. E. Wetton and G. Williams, *Trans. Faraday Soc.*, **61**, 2132 (1965).
9. E. L. Gal'perin, B. P. Kosmynin, L. A. Aslanyan, M. P. Mlenik and V. K. Smirnov, *Vysokomol. Soyed.*, **A12**, 1654 (1970) (in Russian).
10. C. W. Bunn and D. R. Holmes, *Discuss. Faraday. Soc.*, No. 25, 95 (1958).
11. K. Imada, T. Miyakawa, Y. Chatani, H. Tadokoro and S. Murahashi, *Makromol. Chem.*, **83**, 113 (1965).
12. T. Saegusa, H. Imai and J. Furukawa, *Makromol. Chem.*, **56**, 55 (1962).
13. M. Cesari, G. Perego and A. Mazzei, *Makromol. Chem.*, **83**, 106 (1965).
14. J. H. Binks and M. B. Huglin, *Makromol. Chem.*, **93**, 268 (1966).
15. S. Kobayashi, K. Murakami, Y. Chatani and H. Tadokoro, *J. Polym. Sci.: Polym. Lett. Ed.*, **14**, 591 (1976).
16. J. Furukawa and T. Saegusa, "Polymerization of Aldehydes and Oxides", Interscience, New York, NY, p. 225 (1963).
17. A. C. Farthing, in "High Polymers", Vol. 13, N. G. Gaylord (ed.), Interscience, New York, NY, Chap. 5 (1963).
18. M. B. Huglin, D. H. Whitehurst and D. Sims, *J. Appl. Polym. Sci.*, **12**, 1889 (1968).
19. W. R. Sorenson and T. W. Campbell, "Preparative Methods of Polymer Chemistry", Interscience, New York, NY, p. 255 (1961).
20. P. Dreyfuss and M. P. Dreyfuss, Unpublished results.
21. M. P. Dreyfuss and P. Dreyfuss, *J. Polym. Sci*: *A1*, **4**, 2179 (1966).
22. G. M. Palyutkin, A. G. Okuneva, B. V. Vasil'ev and O. G. Tarakanov, *Plast. Massy*, **7**, 67 (1973) (in Russian); through *CA*, **79**, 146920c (1973).
23. T. Kijima, K. Imada and M. Takayanagi, *Reports on Progress in Polym. Phys. Japan*, **16**, 229 (1973).
24. M. Girolamo, A. Keller, K. Miyasaka and N. Overbergh, *J. Polym. Sci.: Polym. Phys. Ed.*, **14**, 39 (1976).
25. V. P. Rushchupkin, B. V. Ozerkovskii and G. V. Korolev, *Vysokomol. Soyed.*, **A11**, 1356 (1969) (in Russian). Translated in *Polym. Sci. USSR*, **11**, 1540 (1969).

26. P. J. Flory, "Principles of Polymer Chemistry", Cornell University Press, Ithaca, New York, p. 570 (1953).
27. B. V. Lebedev and V. Ya Lityagov, *Zh. Fiz. Khim*, **51**(5), 1279 (1977) (in Russian); through *CA*, **87**, 74363p (1977).
28. S. Yoshida, H. Suga and S. Seki, *Polym. J.*, **5**, 25 (1973).
29. Y. Stejny, *J. Macromol. Sci.-Chem.*, **A7**, 1435 (1973).
30. J. Lehmann and P. Dreyfuss, *Adv. Chem. Series*, **176**, 587 (1979).
31. A. H. Willbourn, *Trans. Faraday Soc.*, **54**, 717 (1958).
32. J. Brandrup and E. H. Immergut, "Polymer Handbook", Interscience, New York, p. III-80 (1966).
33. R. E. Wetton, D. B. James and W. Whiting, *J. Polym. Sci.: Polym. Lett. Ed.*, **14**, 577 (1976).
34. D. B. James, R. E. Wetton and D. S. Brown, *Polymer*, **20**, 187 (1979).
35. M. P. Dreyfuss and J. Hugh Macey, *ACS Polym. Preprints*, **20**(2), 324 (1979).
36. A. Davis and J. H. Golden, *Macromol. Chem.*, **81**, 38 (1965).
37. M. B. Neiman, B. M. Kavarskaya, I. I. Levantovskaya and M. P. Yazvikova, *Plast. Massy*, **1**, 42 (1966) (in Russian); through *CA*, **64**, 12827h (1966).
38. P. A. Okunev, *Sb. Nauch. Tr. Ivanov. Energ. Inst.*, **14**, 130 (1972) (in Russian); through *CA*, **81**, 64507a (1974).
39. J. H. Golden, *Makromol. Chem.*, **81**, 51 (1965).
40. C. Thomassin and J. Marchal, *Makromol. Chem.*, **178**, 981 (1977).
41. C. Thomassin and J. Marchal, *Makromol. Chem.*, **178**, 1295 (1977).
42. C. Thomassin and J. Marchal, *Makromol. Chem.*, **178**, 1313 (1977).
43. C. Thomassin and J. Marchal, *Makromol. Chem.*, **178**, 1327 (1977).
44. P. J. Hendra and H. A. Majid, *Polymer*, **18**, 573 (1977).
45. J. F. Rabolt, *J. Polym. Sci.: Polym. Phys. Ed.*, **17**, 1457 (1979).
46. A. Tsutsumi, K. Hikichi and M. Kaneko, *Polym. J.*, **8**, 443 (1976).
47. M. Okada, Y. Yokoyama and H. Sumitomo, *Makromol. Chem.*, **170**, 131 (1973).
48. Y. Tsujita, T. Nose and T. Hata, *Polym. J.*, **5**, 201 (1973).
49. Y. Tsujita, T. Nose and T. Hata, *Polym. J.*, **6**, 51 (1974).
50. D. Sims, *Polymer*, **6**, 220 (1965).
51. B. V. Lebedev and V. Ya. Lityagov, *Vysokomol. Soyed.*, **A19**, 2283 (1977) (in Russian). Translated in *Polym. Sci. USSR*, **19**, 2618 (1977).
52. G. A. Clegg, D. R. Gee, T. P. Melia and A. Tyson, *Polymer*, **9**, 501 (1968).
53. R. C. Burrows, *ACS Polym. Preprints*, **6**(2), 600 (1965).
54. G. G. Winspear, "The Vanderbilt Rubber Handbook", R. T. Vanderbilt Co., Inc., New York, NY (1958).
55. "Compounders Technical Guide", CTG No. 49, Vol. 1, The B. F. Goodrich Co., Research Center, Brecksville, Ohio, (undated).
56. C. D. Hodgman, "Handbook of Chemistry and Physics", The Chemical Rubber Publishing Co., Cleveland, Ohio (1960).

57. R. E. Wetton and G. Allen, *Polymer*, **7**, 331 (1966).
58. J. M. Perena and J. M. G. Fatou, *An. Fis.*, **73**(4), 267 (1977) (in Spanish); through *CA*, **91**, 5614s (1979).
59. I. Penczek and Z. Kohman, *Polimery*, **15**(2), 63 (1970) (in Polish); through *CA*, **73**, 46403u (1970).
60. A. I. Marei, E. A. Sidorovich, G. E. Novikova and E. I. Rodina, *Vysokomol. Soyed.*, **A10**, 542 (1968) (in Russian); through *CA*, **69**, 11225c (1968).
61. Goodyear Tire and Rubber Co., UK Patent 1,006,316 (Cl.C 08g) (Sept. 29, 1965).
62. Goodyear Tire and Rubber Co., UK Patent 1,001,345 (Cl.C 08g) (Aug. 18, 1965).
63. A. P. Dunlop, N. E. Rustad and E. Sherman, US Patent 3,544,487 (to Quaker Oats Co.) (Dec. 1, 1970).
64. K. Bak, G. Elefante and J. E. Mark, *J. Phys. Chem.*, **71**, 4007 (1967).
65. Yu. A. Gorin, E. I. Rodina, N. V. Kozlova and K. V. Nel'son, *Vysokomol. Soyed.*, **A11**, 1477 (1969) (in Russian). Translated in *Polym. Sci. USSR*, **11**, 1675 (1969).
66. I. Kuntz and R. N. Thomas, US Patent 3,435,015 (to Esso Research and Engineering Co.) (Mar. 25, 1969); *CA*, **70**, 10756p (1969).
67. I. Kuntz, *Trans. NY Acad. Sci.*, **33**, 529 (1971).
68. H. Steppan, G. Messwarb, F. Erdmann and W. Lueders, German Patent 1,745,044 (Cl. CO8G) (to Hoechst A.-G.) (Sept. 18, 1975).
69. G. G. Winspear, "The Vanderbilt Rubber Handbook", R. T. Vanderbilt Co., New York, NY (1968).
70. S. Smith, A. J. Hubin, J. J. Hoffman and D. D. Zimmerman, *Am. Chem. Soc., Div. Rubber Chem., Preprints*, No. 43 (1967).
71. G. H. Brice, US Patent 3,373,143 (to Uniroyal, Inc.) (Mar. 12, 1968).
72. P. Dreyfuss, M. P. Dreyfuss and H. A. Tucker, *Adv. Chem. Series*, **128**, 125 (1973).
73. R. Block, O. Kedem and D. Vofsi, *J. Polym. Sci.*, **B3**, 965 (1965).
74. S. Ono, K. Koyama and A. Shimizu, Japan. Kokai 78 16,373 (to Toyo Soda Mfg. Co., Ltd.) (Feb. 15, 1978); through *CA* **89**, 77017f (1978).
75. S. Ono, K. Koyama and A. Shimizu, Japan. Kokai 78 28,661 (to Toyo Soda Mfg. Co., Ltd.) (Mar. 17, 1978); through *CA*, **89**, 60695c (1978).

CHAPTER 7

Industrial Applications: Polyurethane and Polyester Thermoplastic Elastomers

PTHF has many desirable properties which have been utilized by industry in polyurethane and polyester thermoplastic elastomers. In these applications the tendency of PTHF to crystallize slowly at room temperature is overcome by copolymerization, and difficulties in crosslinking are circumvented by the introduction of copolymeric segments capable of forming networks. As a homopolymer, PTHF is currently too expensive to complete with general-purpose or even specialty elastomers; but, as the soft segment of thermoplastic elastomers, PTHF is widely used industrially. In these applications it is valued as a precursor leading to elastomers with outstanding hydrolytic stability, good low-temperature flexibility, outstanding thermal stability at elevated temperatures and high fungus resistance, combined with excellent abrasion resistance, resiliency and dynamic properties.[1–5]

Thermoplastic elastomers are not necessarily chemically crosslinked, although they may be in the final processing step. Instead they behave as rubbers because of their unique morphology. Chemically they are composed of blocks of so-called "soft" and "hard" segments. The "soft" segment is an amorphous polymer above its glass-transition temperature. The "hard" segment is a polymer below its glass-transition temperature that is capable of undergoing thermally reversible intermolecular associations. These segments do not mix uniformly. Instead they segregate and result in domain formation. The intermolecular associations within the domains function as pseudocrosslinks to provide dimensional stability and recovery after mechanical deflection. Above a certain temperature the "hard" segments soften and the copolymers can be processed in conventional plastics equipment such as extruders and injection-molding machines.[5]

Three types of molecular structure can lead to thermoplastic elastomer characteristics:

a) ABA triblocks in which an elastomer block, B, is capped on both ends by a hard block, A;

b) graft copolymers in which the backbone is an elastomer and the branches are hard segments or vice versa;

c) segmented molecules in which hard and soft blocks are randomly distributed in the backbone itself.

ABA triblocks with PTHF as the B block and graft copolymers containing PTHF have been prepared but, as far as it has been possible to establish, none has been commercialized. Their synthesis and properties will be discussed in Chapter 8. This chapter will deal with two types of segmented molecules containing PTHF that have important commercial markets: the first type are polyurethanes in which the intermolecular associations are hydrogen bonds, while the second are polyesters with hard segments that undergo crystallization. Glass formation is another mechanism by which the hard segments can associate, but this mechanism is unimportant in PTHF segmented copolymers.

7.1 SYNTHESIS

Commercial polyurethanes and polyesters containing PTHF are based on α, ω-difunctional precursors of relatively low molecular weight. Typically, these precursors are waxy solids with number-average molecular weights ranging from 650 to 3,000. Most often diprimary diols are used, but for some polyurethane applications it is advantageous to convert the hydroxyl groups to isocyanate groups first.

7.1.1 Diprimary diols

In the US, diprimary diols from PTHF are available in commercial quantities from du Pont and Quaker Oats under the trade names "Teracol Polyether Glycols" and "Polymeg Polyols", respectively. In 1973 annual consumption of such glycols was about 30 million pounds in the US and about five million pounds in Japan.[6]

Numerous names have been used for these glycols. Among them are poly(tetramethylene ether) glycol (PTMEG), poly(1,4-oxybutylene) glycol, poly(oxytetramethylene) (PTOM) and α-hydro-ω-hydroxypoly(oxytetramethylene). The parentheses are sometimes omitted, especially in older literature. In this chapter we will again follow *Chemical Abstracts* and use poly(tetramethylene ether) glycol. PTMEG, diprimary glycols from PTHF, PTHF glycol, polyols and diprimary diols of PTHF will also be used.

Many methods of preparing PTMEG can easily be imagined after a consideration of the chemistry outlined in Chapter 3. Only a few of these are important commercially. Nevertheless, a variety of methods will be mentioned here to give the reader some idea of the chemistry that is potentially available.

Protonic-acid initiation The simplest and most direct method of preparing PTMEG is to initiate the polymerization of THF with protonic acid to produce an OH head group and to terminate the polymerization with water to produce an OH end group.

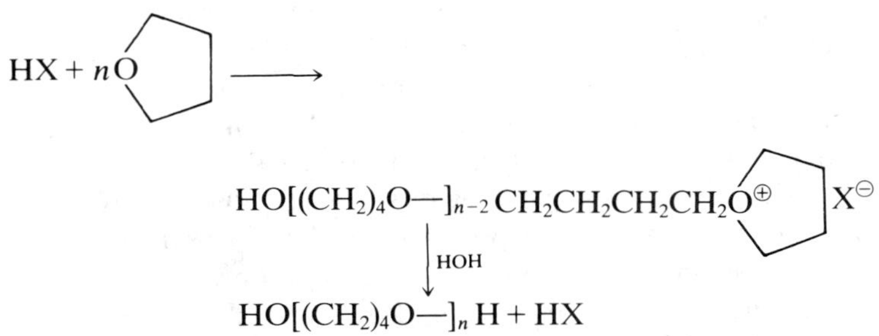

This is the first and traditional method that has been used. It has been reported that polymerizing THF with fluorosulfonic acid readily yields PTHF with hydroxyl groups at both ends after treatment with water.[7] Sometimes SO_3 [6] and/or $HClO_4$ [7] is added to the principal acid initiation. Thus, initiation with H_2SO_4 containing 15–45 wt.% SO_3 and termination with water is reported to give diprimary diols.[6,8]

These processes have several disadvantages. The acids cannot be recovered and reused. Disposal of the acidic byproducts is a problem, because of their toxicity and corrosiveness. Strong acids can cause degradation of the PTMEG. Moreover, as already noted in Chapters 3 and 6, the presence of hydroxyl groups during polymerization markedly increases the molecular weight and broadens the molecular-weight distribution.[9]

Diacetate esters of poly(tetramethylene ether) glycol Many of the problems associated with the direct preparation of PTMEG can be overcome by preparation of a diacetate-ester intermediate. This is accomplished by polymerization of THF using strong protonic acid in the presence of acetic anhydride.[10–14] Typical combinations include perchloric acid–acetic anhydride[10–12] and hexafluoroantimonic acid–acetic anhydride.[13] The first step in the polymerization of THF in these processes is probably protonation of THF to give a secondary oxonium ion. But, in the presence of acetic anhydride, the hydroxyl would be quickly converted to an acetate. Also transfer reactions with acetic anhydride lead to stable acetate end groups and molecular-weight control, in much the same manner as already described for trimethylorthoformate, which gives two methoxy end groups.[15–16] The diprimary glycol can be obtained from the diacetate ester by alcoholysis[17] or hydrolysis of the diacetate by boiling for two to three hours with 0.5 N alcoholic KOH. PTHF glycols with molecular weights up to about 10,000 can be purified by dissolving in ether and washing with water. The purified polymer can be isolated by drying over, for example, sodium sulfate, evaporating the ether and drying under vacuum at 50–60°C.[12]

A very interesting strong-acid catalyst that du Pont may use commercially for the synthesis of hydroxy-terminated polymers and copolymers of THF is Nafion resin ($NfSO_3H$).[17–20] In its free acid form this resin has the structure:[21]

$$—[(CF_2CF_2)_x(CF_2\underset{\substack{|\\(OCF_2\underset{\substack{|\\CF_3}}{C}F)_z—OCF_2CF_2SO_2OH}}{C}F)_y—]_n$$

When PTHF glycols are prepared using this resin the reaction mixture normally includes acetic anhydride and acetic acid in addition to THF.[21] Other anhydrides, such as propionic anhydride, and other acids, such as formic acid, can be used. As shown

TABLE 7.1
Control of Molecular Weight[21] of PTHF Diacetates[a]

Weight of Ac_2O, g	Weight of AcOH, g	Ac_2O/AcOH[b]	Pzn time, hours	Conversion, %	$\overline{M}_n$
3.1	3.1	1:1	1.5	37	1528
3.1	3.1	1:1	5.5	60	1488
3.1	3.1	1:1	21	61	1397
5.6	0.6	9:1	1.5	47	2808
5.6	0.6	9:1	5.5	74	1270
5.6	0.6	9:1	21	82	1002
0.6	5.6	1:9	21	23	2304

[a]All charges contained 55.8 g of THF.
[b]Weight ratio.

in Table 7.1, the ratio of anhydride to acid serves as a means of controlling the molecular weight of the resulting diacetates. Many reactions probably occur simultaneously and in a competing manner in the course of this synthesis;[22] for example, clearly,

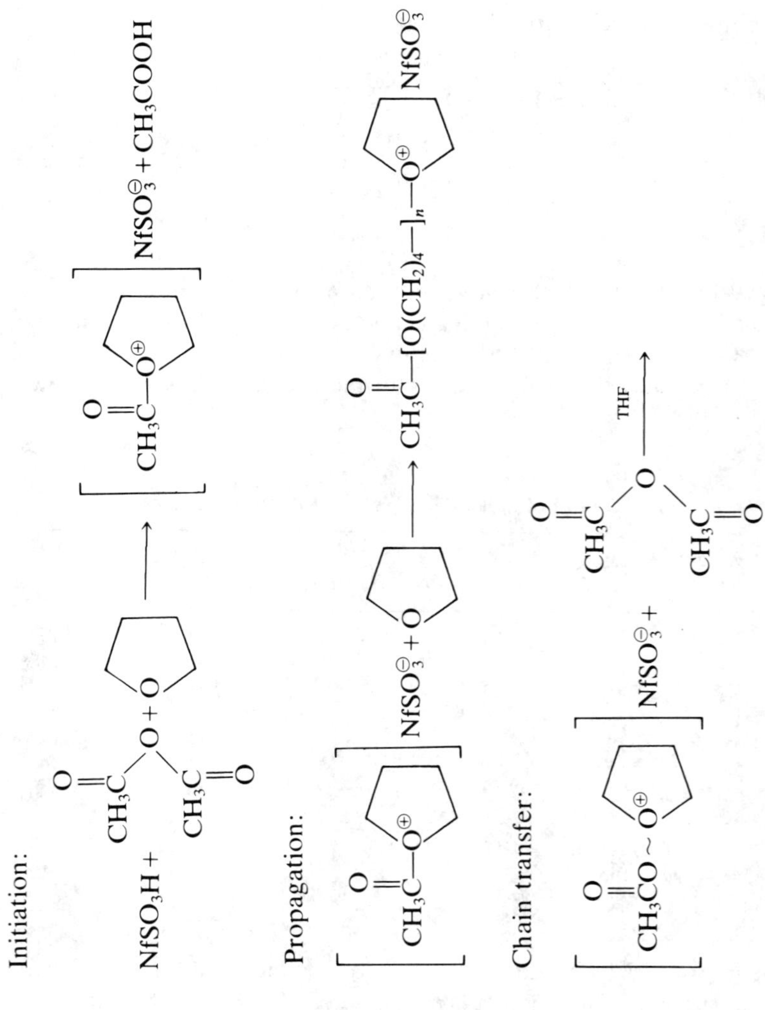
Initiation:
Propagation:
Chain transfer:
THF

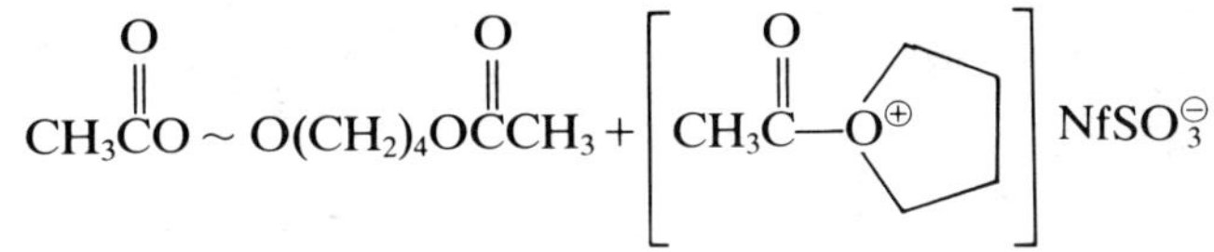

Redistribution:

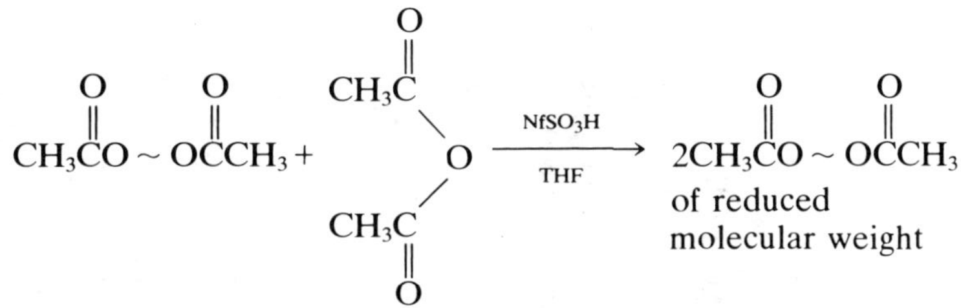

Termination:

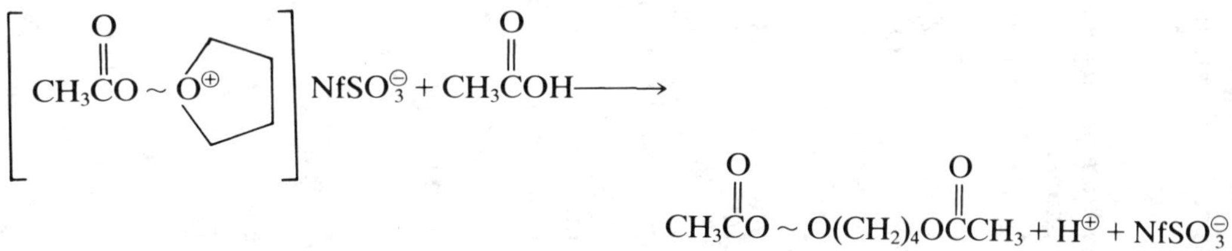

similar reactions would occur with HSO_3F–acetic anhydride–acetic acid mixtures, except that free acid would be one of the products of the termination step.

Difunctional initiators Some initiators lead to PTHF with oxonium ions on both ends of the chain. When these difunctional PTHFs are treated with water, PTMEG is formed:

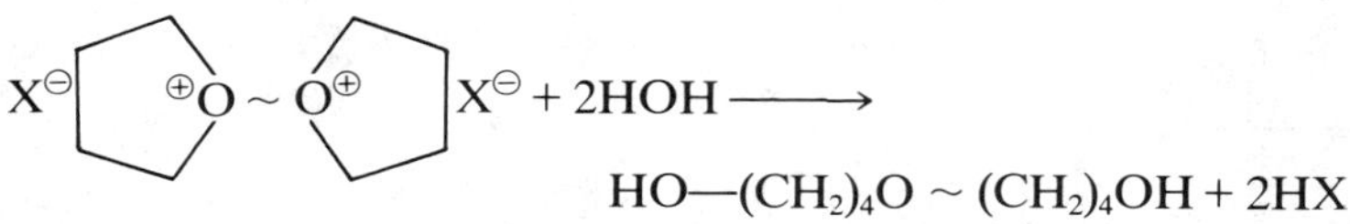

These initiators are few in number and will be presented in their order of appearance in the open literature.

2,2′-Octamethylene-bis-1,3-dioxolenium perchlorate The synthesis of this salt (hereafter called the *bis*-dioxolenium salt) as described by Yamashita and coworkers[23] is lengthy, and its initiating ability is rather slow, but it does lead to a bifunctional PTHF.[23,24]

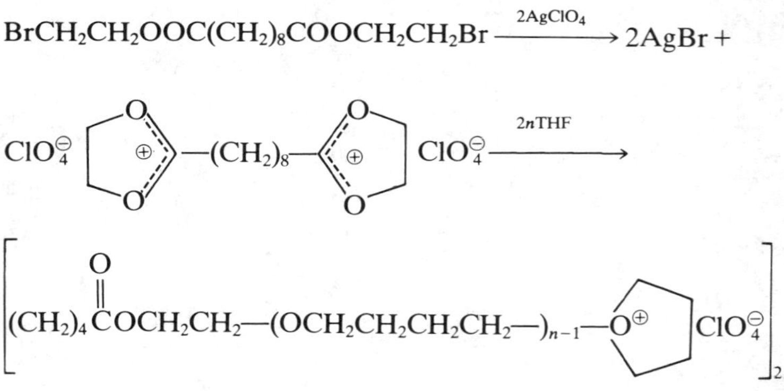

Anhydride of trifluoromethanesulfonic acid A simpler synthesis of difunctional PTHF was reported by Smith and Hubin.[25] The initiator they used is the anhydride of a very strong acid.

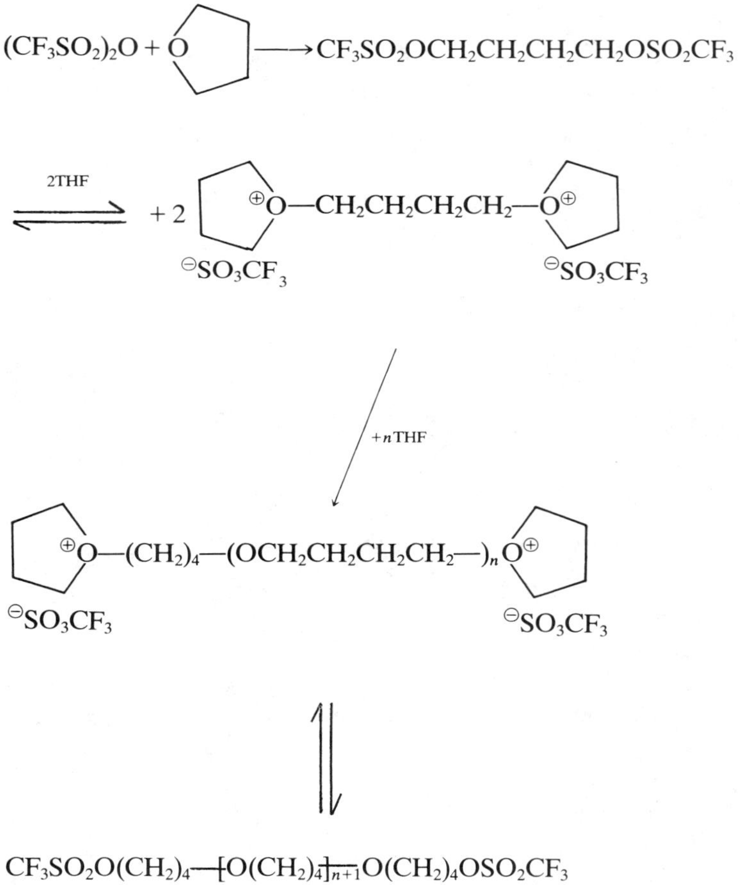

Disulfuryl fluoride, $(FSO_2)_2O$, behaves similarly.

Use of the anhydride eliminates the complications resulting from the presence of hydroxyl groups in polymerizations initiated by HSO_3F, but is otherwise similar.

Phosphorus pentafluoride Hoene and Reichert have studied the initiation step in the cationic polymerization of THF initiated by PF_5 by ^{31}P NMR.[26] They concluded that bifunctional PTHF is formed.

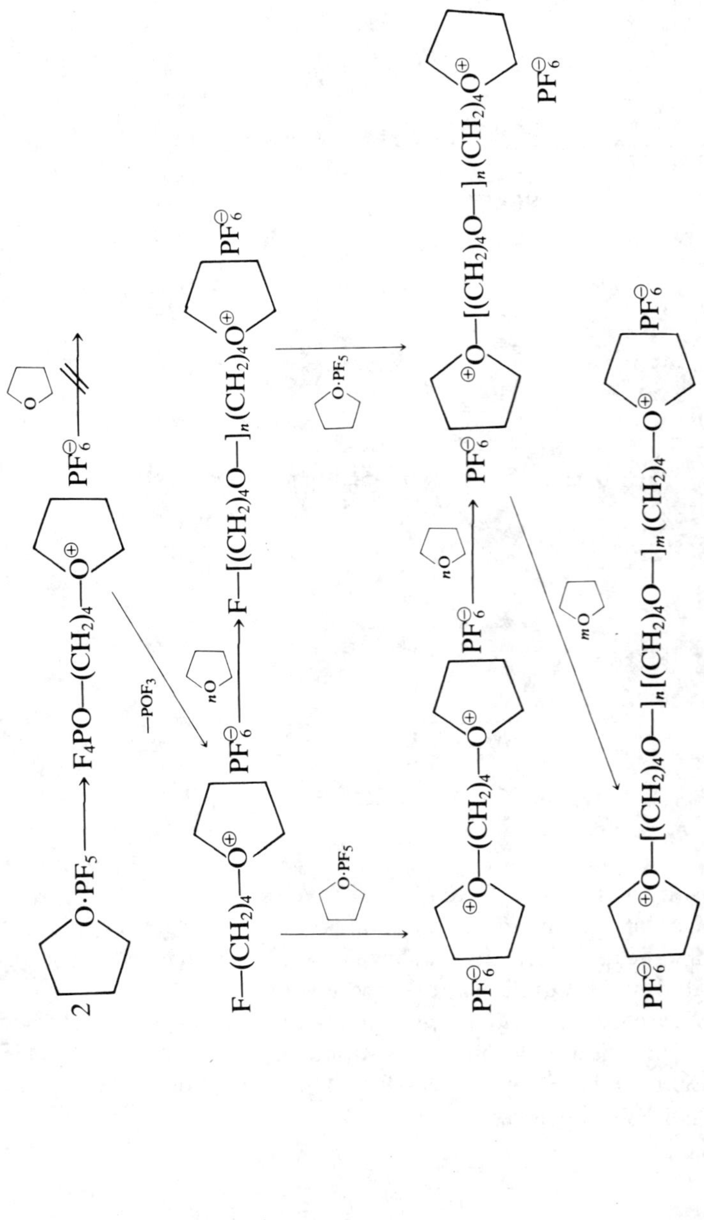

Terephthaloyl and adipoyl salts Lehmann and co-workers[27,28] have polymerized THF using diacid chlorides and silver salts or preformed acyl salts having complex counter-ions. They have shown by a variety of molecular-weight methods that initiation is very efficient, nearly 100%.

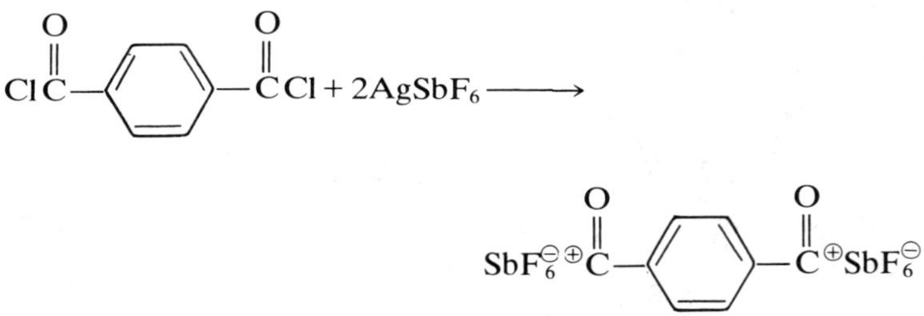

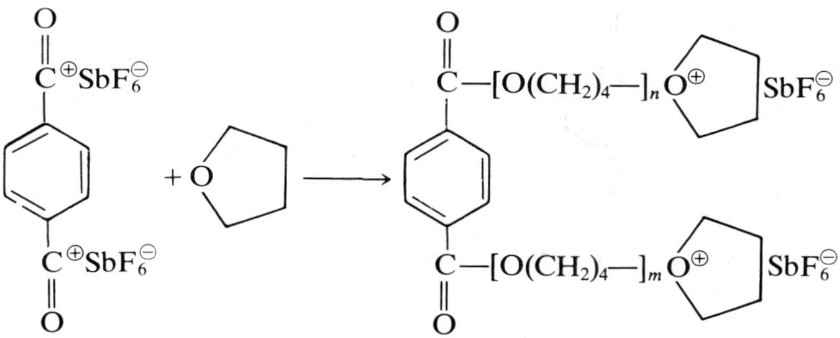

Dialkylhalides and silver salts Acyl halides are more reactive than alkyl halides. Usually the intermediate salts cannot be isolated when alkyl halides are used.[30] However, difunctional PTHF has been prepared from xylylenedibromide and silver perchlorate or silver hexafluorophosphate.[29]

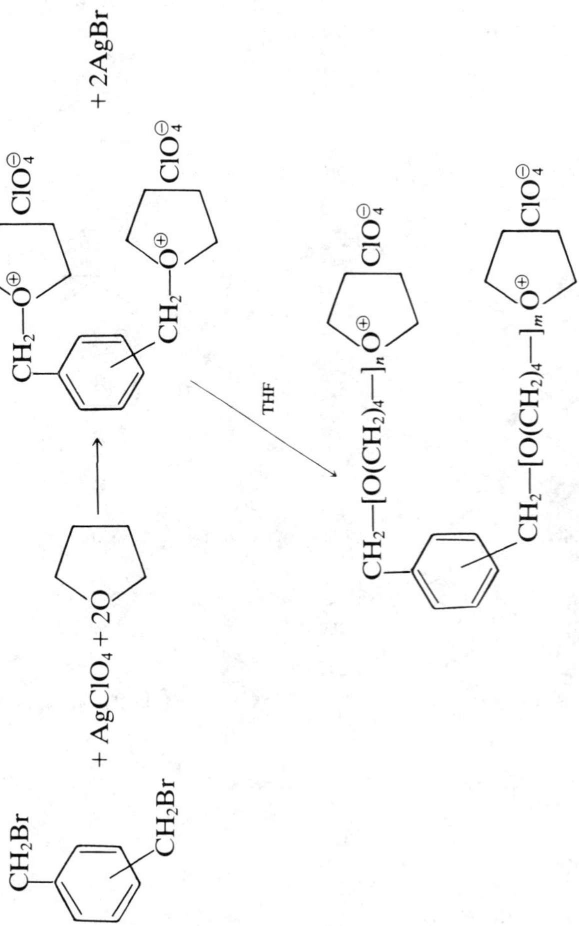
CH2Br
CH2Br
+ AgClO4 + 2O
CH2—O
ClO4
+ 2AgBr
THF
CH2—[O(CH2)4—]n O
CH2—[O(CH2)4—]m O
ClO4

Silenium dicationic initiator Difunctional PTHF has also been prepared from the silenium dicationic initiator below by Kucera *et al.*,[31,32] who report that the rate of initiation is high and that the conversion of PTHF is directly proportional to time.[31]

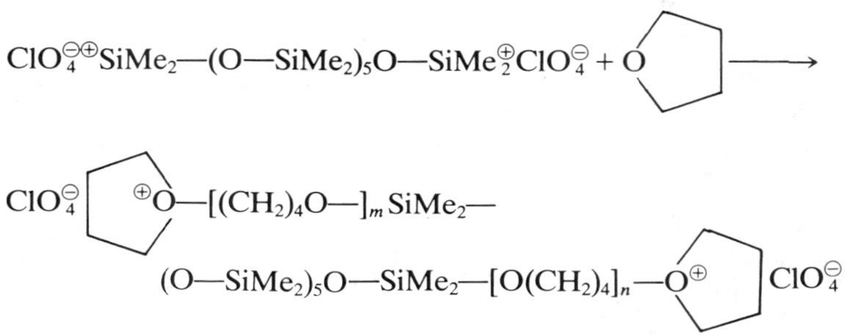

The data in the literature presently available are not adequate to specify whether $m = n$, but it may be.

7.1.2 Precursors with isocyanate end groups

Diisocyanates are readily prepared from the dry diprimary glycols by reaction under nitrogen with diisocyanates. The compound most commonly used is 2,4-toluene diisocyanate (TDI),[33] but good products have been prepared from methylene-*bis*(4-cyclohexyl isocyanate)[34] or diphenylmethane-p,p′-diisocyanate (MDI).[35] Linear isocyanate compounds are produced. Typically, dry glycol is added to the diisocyanate at such a rate that the temperature does not exceed 50–55°C. After the exotherm ceases, the reaction mixture is heated to 80°C and maintained at that temperature for one to two hours to complete the reaction.[33] The percentage of free isocyanate in the resulting product depends primarily on the molecular weight of the original polyol and the mole ratio of TDI to polyol used.

Isocyanate-terminated precursor liquids are available commercially with free isocyanate groups in the range 3–9 wt.%. In the US, du Pont markets a family of these materials under the

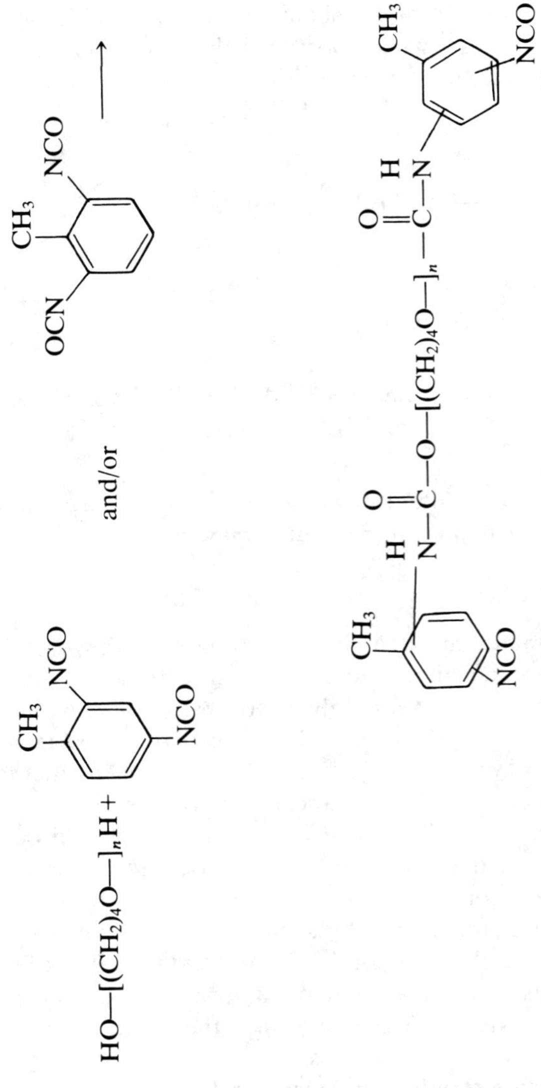
HO—[(CH2)4O—]nH +
CH3
NCO
NCO
and/or
OCN
CH3
NCO
CH3
NCO
H
N—C—O—[(CH2)4O—]n—C—N
O
O
H
CH3
NCO

trade name "Adiprene L", while Uniroyal sells "Vibrathanes".

7.1.3 Polyurethanes

Polyurethanes can be prepared from either the diprimary diols or from the isocyanate-terminated precursors. Diols are normally converted to isocyanate-terminated prepolymers before use, but polyurethanes with excellent physical properties and good processability can also be prepared by combining the polyol, an aliphatic glycol and a diisocyanate such as MDI.[34] If the ratio of the components is controlled so that the amounts of combined glycols and the amount of diisocyanate are equal, no further vulcanization step is needed. Thus, two methods are available to produce urethane elastomers from the polyols: namely, prepolymer and one-shot. The prepolymer method, in which a diisocyanate is first prepared and then extended, is usually preferred. It permits greater control of the individual reactions than when all reactants are added at once.

Chain extenders The precursors can be extended with monomeric diols, diamines or a mixture of diols and diamines. The most important chain extenders include 1,4-butanediol and the diamines used for curing. The amount and nature of chain extender used influences the final properties.[35,36] Some typical recipes will be given in the section on properties of polyurethanes.

Curing The chain-extended polymers are typically cured by reaction of the isocyanate groups with monomeric diamines or triols. Curing can be carried out over a wide range of temperatures, although temperatures below 116°C are preferable. This produces vulcanizates with ordered structures and optimum properties. Curing above 116°C increases the amount of branching due to biuret formation and decreases physical properties. To achieve this curing, temperature control, mixing and degassing are carried out at 70–100°C; in this temperature range, the precursors have good stability and low viscosity and may be degassed readily. Curing is an exothermic reaction and can lead to an increase by as much as 25°C in the temperature of the reaction mass. Curing time can be varied from five minutes to 24 hours, depending upon the curing agent and temperature. Pot life similarly varies from 1 to 15 min depending on the recipe.

Curing with amines A good general-purpose curing agent is 4,4′-methylene-*bis*(2-chloroaniline) (MBCA).[35,36]

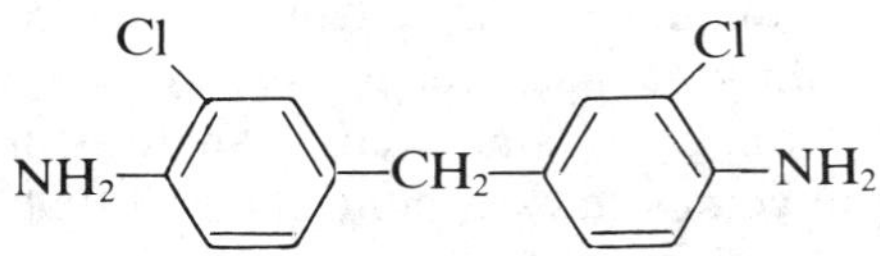

Methylene dianiline (MDA), $NH_2C_6H_4CH_2C_6H_4NH_2$ and mixtures of diamines such as *m*-phenylene diamine and cumene diamine or MBCA and MDA or even ternary mixtures of diamines can be used for curing. When mixtures are used, processing characteristics and vulcanizate properties lie somewhere between those obtained with the diamines individually.[35,36]

Acid catalysts such as adipic, acetic, benzoic or oleic acids can be used to accelerate the curing process in the presence of diamines.[35,36]

Curing with triols Triols such as trimethylolpropane, $C_2H_5C(CH_2OH)_3$, are also satisfactory curing agents. Triols are normally used in combination with diols as chain extenders. Triethylamine, stannous octoate or ferric acetyl acetonate at levels of 0.0005 to 0.5 parts per hundred of diisocyanate terminated precursor significantly increase cure rates.[35,36]

Choice of materials The materials named above are not the only materials that can be used in the preparation of polyurethanes from PTHF glycols. They are intended to be illustrative of the scope of the methods of synthesis. Choice of a specific recipe and reaction conditions is determined by the properties and processing characteristics required for a given application. For further details, the reader should consult the open literature and company bulletins.[33,35–40]

Chemistry The chemistry involved in the preparation of polyurethanes has been reviewed many times[37–39] and only the basic reactions will be presented here. These include the following: a) *Capping with isocyanate endgroups*: *urethane formation*

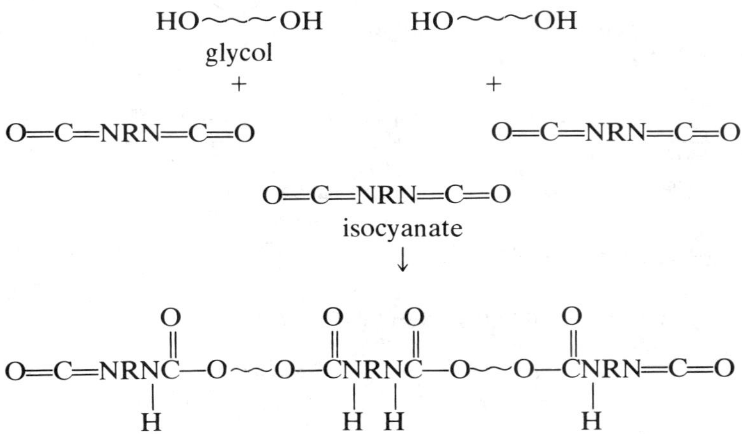

urethane (isocyanate-terminated prepolymer)

b) *Chain extension with monomeric glycol*: *urethane formation*

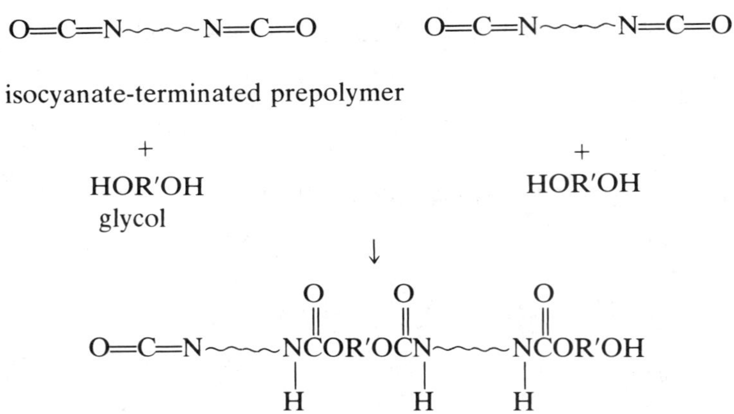

urethane

c) *Crosslinking/branching*: *allophonate formation*

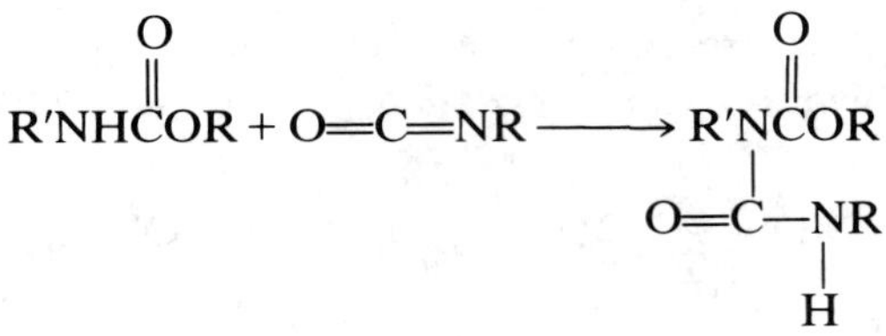

urethane isocyanate allophonate

d) *Chain extension with diamine*: *urea formation*

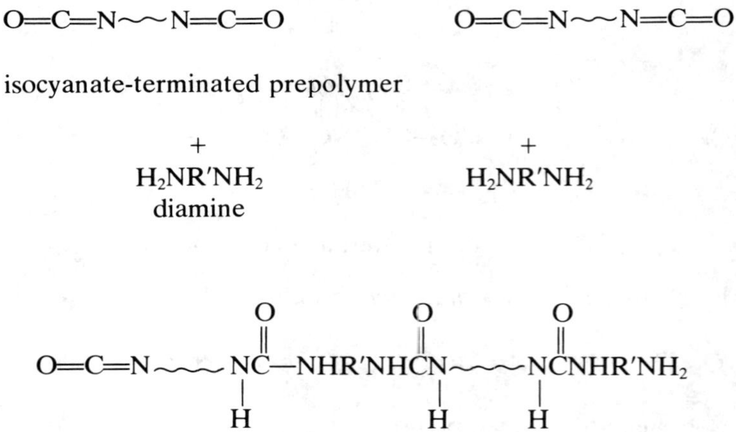

e) *Crosslinking/branching*: *biuret formation*

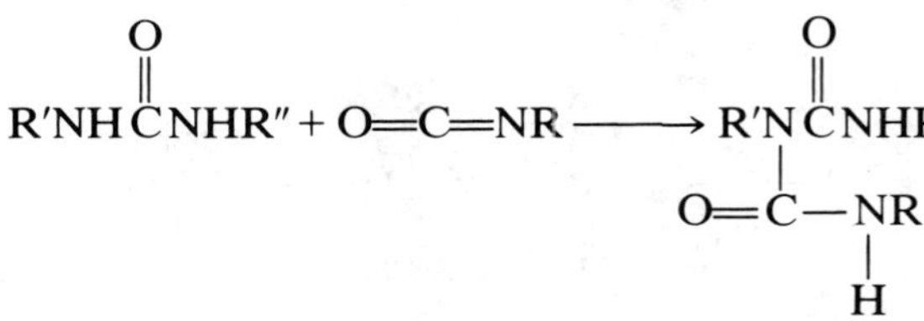

urea isocyanate biuret

Reactions of isocyanates with urethanes and with ureas to give allophonates and biurets, respectively, are important because, in the absence of added triols, they are the means by which branching and crosslinking occur in polyurethane synthesis. A slight excess of diisocyanate is normally employed for this purpose. These reactions are slow compared to urethane formation. But urethane formation occurs at a moderate rate compared to urea formation; therefore, the isocyanate-terminated prepolymer is preferably formed before the final curing reaction with diamine.[39]

Reaction with water Isocyanates also react with water to give CO_2 and amine

$$HOH + O{=}C{=}NR \longrightarrow [HO{-}\overset{\displaystyle O}{\overset{\|}{C}}\underset{\displaystyle H}{\underset{|}{N}}R] \longrightarrow CO_2 + H_2NR$$

Since urea formation is very fast, water must be excluded during the formation of the prepolymer to ensure control of the product and its properties. Alternatively, water can be used as a curing additive in some systems.

7.1.4 Polyesters

Commercial block copolymers are condensation products of aromatic dicarboxylic acid, low-molecular-weight PTMEG or its copolymers and monomeric diols. They are prepared by melt polymerization procedures using starting materials like dimethyl esters of terephthalic acid, glycols of 1,000 molecular weight and various monomeric glycols, especially, 1,4-butanediol, ethylene glycol[5,41–43] and 1,4-cyclohexanedimethanol.[44,45]

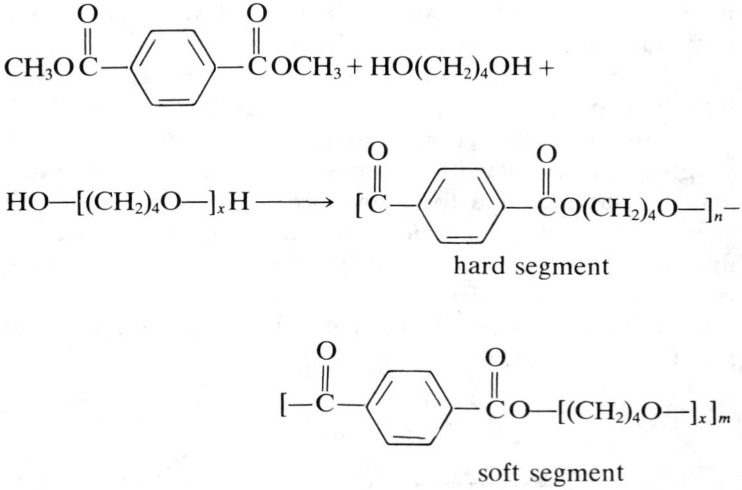

Mixtures of dicarboxylic acids and occasionally copolymers of THF are used.

As usual, a prepolymer with ~50% unreacted hydroxyl groups is prepared by interchange of the methyl ester of the aromatic dicarboxylic acid with a mixture of polymeric and monomeric diols. A titanate transesterification catalyst is frequently used. Reaction mixtures usually contain up to 1% of an aromatic amine or hindered phenol antioxidant, which is added at the beginning of the reaction. Methanol is fractionated off the reaction mixture, and then the prepolymer is driven to high molecular weight by distilling off the excess monomeric diol at 240–50°C and a pressure of less than 1 torr.[5,42]

The method of polymerization has some limitations. The melt must remain homogeneous during polymerization. Hence, long reaction time or higher reaction temperatures may be required for very-high-melting hard segments or sterically hindered acids or glycols. These relatively severe conditions tend to limit the molecular weight and hence the physical properties that can be attained.

7.2 PROPERTIES AND USES

To date, diprimary diols from PTHF and indeed PTHF itself find major applications only in polyurethanes, although applications of polyester thermoplastic elastomers from PTHF are increasing.

Two different processing technologies are used with polyurethanes; namely, thermoplastic and thermoset. The major advantages of thermoplastic compounds are their ability to be rapidly molded and reprocessed. Generally, systems derived from the diprimary diols, MDI, and 1,4-butanediol provide the required property and processing characteristics for thermoplastics.[1]

The diols can be converted to highly crosslinked cast elastomers using both TDI and MDI thermoset technologies. In TDI–diamine systems, the linearity of the diols permits a high degree of chain entanglement, resulting in high-performance elastomer products.

Elastomers based on the glycols from PTHF have excellent fungal resistance, provide good tensile strength, tear and abrasion resistance and exhibit high vapor transmission and low hysteresis loss. Properly purified products are approved by the US Federal Drug Administration (FDA) for use in adhesives[46] and polyurethanes which come into direct contact with dry food.[47]

As urethanes, the glycols have many industrial applications. For thermoplastics these include cable jacketing, tubing, sheet and film, rollers, flexible couplings, cattle name tags, gears, conveyor belt facings, pump impellers, fasteners, prosthetic devices, O-rings, diaphragms and skateboard wheels.[1] Examples of uses for thermosets are ski boots, tires, caster wheels, rollers, gasketing, press brakes, gears, auto belts, mine belting, mallets and hammers, bushings and bearings, cable guides, encapsulating and potting compounds, impact mountings and tooling.[1] Other applications of these urethanes are in coatings for bowling alleys, gymnasiums and marine finishes, elastic films ("Spandex[tm]"), adhesives for bonding rubber, wood, metal fabrics and plastics, upholstery and synthetic leathers.[1]

TABLE 7.2
Typical Properties of[1,2] "Polymeg Polyols"[a] and "Teracol" Polyether Glycols[4]

Property	"Polymeg 1000"	"Polymeg 2000"	"Teracol 650"	"Teracol 1000"	"Teracol 2000"	"Teracol 2900"
Molecular weight	1000 ± 50	2000 ± 100	600 ± 70	1000 ± 50	2000 ± 100	2900 ± 100
Hydroxyl number	107–18	53–59	160–87	107–18	53–59	37–40
Acid No. max.	0.05	0.05	0.05	0.05	0.05	0.05
Moisture, % max.	0.03	0.03	0.03	0.03	0.03	0.03
Volatiles, % max.	0.1	0.1	—	—	—	—
APHA color, max.	90	90	90	90	90	90
Peroxides, % (active O)	0.000	0.000	—	—	—	—
Unsaturation, Meq/g	0.0011	0.0010	—	—	—	—
Ash, ppm	10	10	—	—	—	—
Ca, ppm	3	3	—	—	—	—
Fe, ppm	0.3	0.3	—	—	—	—
Saponification No.	0.00	0.00	—	—	—	—
Melting temperature, °C	38	38	10–32	10–32	10–32	10–32
Solidification point, °C	19	22	—	—	—	—
Boiling point, °C	>204	>204	—	—	—	—
Flash point, open cup, °C	>260	>260	>260	>260	>260	>260
Specific heat, cal./g/°C	0.523	0.503				

[a]"Polymeg 650" with molecular weight 650 ± 50 was discontinued in October 1978.

In other compounds, the glycols are used mainly in polyester resins, film modifiers and polyester thermoplastics. Potential markets include hydraulic hose tubes and covers, chemical hose, low-pressure tires, speciality belting, snowmobile tracks, cable jacketing, seals, coatings and various extruded and molded goods.

This chapter is not intended to provide comprehensive coverage of all the variables affecting the properties of products derived from PTHF. Instead, typical examples have been selected from the literature to illustrate the range of materials that have been prepared. No effort has been made to assemble a comprehensive list of references, although it is hoped that no key references have been omitted, and that a balanced picture has been presented.

Properties of the diols, of precursors with isocyanate end groups, of urethanes and of polyesters will be considered separately. The factors affecting properties will be emphasized.

7.2.1 Diprimary diols

In the US there are two major suppliers of glycols based on PTHF. As stated above, Quaker Oats sells "Polymeg Polyols" and du Pont markets "Teracol Polyether Glycols".

"Polymeg Polyols" are commercially available in two molecular weights: 1,000 and 2,000. Their properties are summarized in Table 7.2.[1,2] The toxicity data given in Table 7.3[1]

TABLE 7.3
Toxicity Data for "Polymeg Polyols"[1]

Test	"Polymeg 1000"	"Polymeg 2000"
Acute oral toxicity, LD_{50}, mg/kg (albino rats)	18,830	>34,000
Acute dermal toxicity, LD_{50}, mg/kg (albino rabbits)	>10,250	>10,250
Eye irritation (albino rabbits)	Minimally irritating	
Primary skin irritation (albino rabbits)	Minimally irritating	
Threshold limit value	None	

show that the glycols are nonhazardous materials. The variations of specific gravity with molecular weight and temperature are shown in Figures 7.1 and 7.2.[1,2] Variations of specific viscosity with the same variables are shown in Figure 7.3.[2] GPC tracings confirm that "Polymeg Polyols" have a relatively

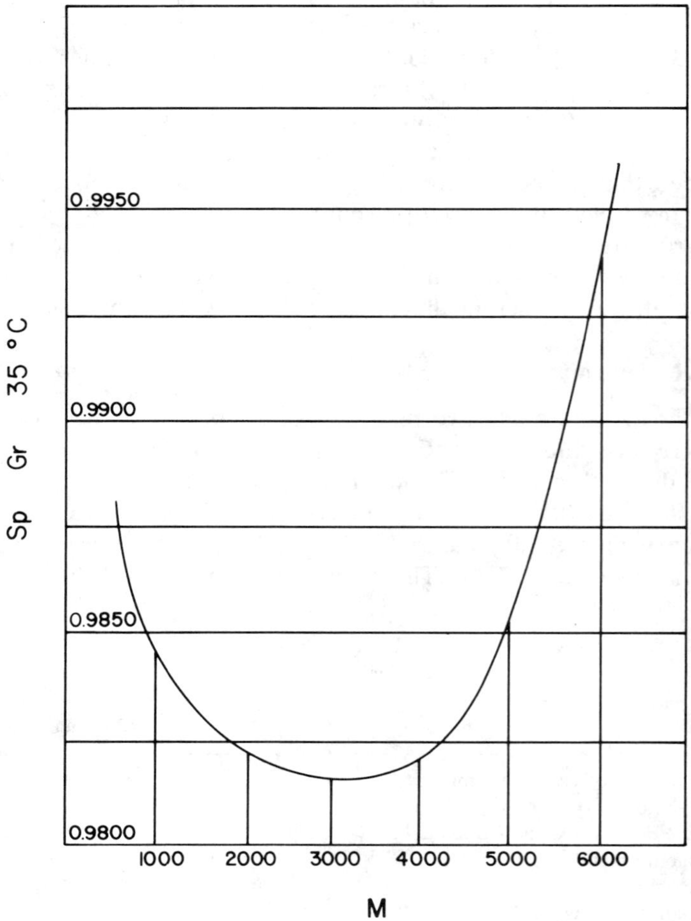

FIGURE 7.1 Specific gravity against molecular weight for "Polymeg Polyols".[1,2]

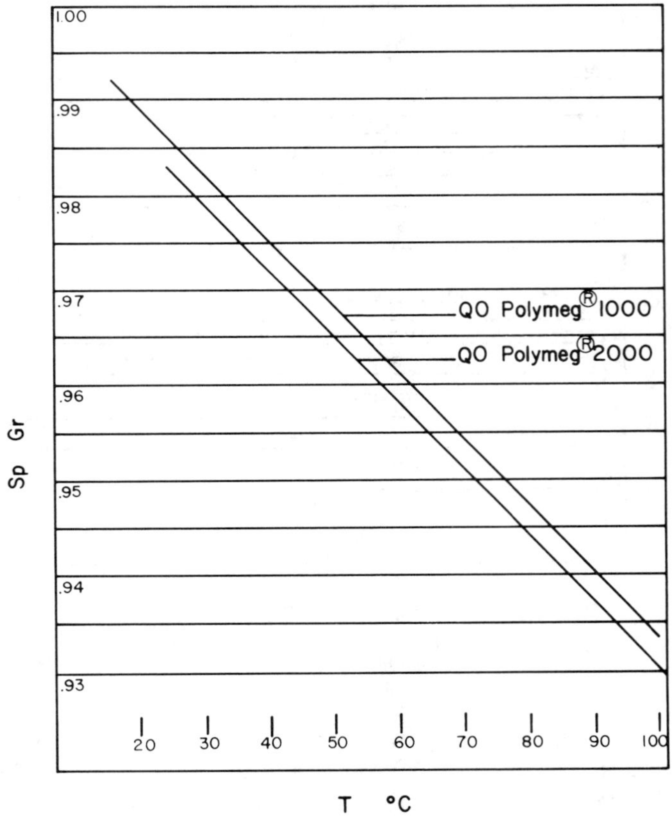

FIGURE 7.2 Specific gravity against temperature for "Polymeg Polyols".[2]

narrow molecular weight distribution and IR spectra show only absorption peaks due to hydroxyl, methylene, tetramethylene and ether linkages. Table 7.2 indicates that "Polymeg Polyols" have 0.0011 meq/g of unsaturation. Even if all the measured unsaturation were in the end groups, the total amount of monofunctionality would be less than 0.1%.

The glycols are soluble in aromatic and chlorinated hydrocarbons, alcohols, esters, ketones and nitroparaffins. They are

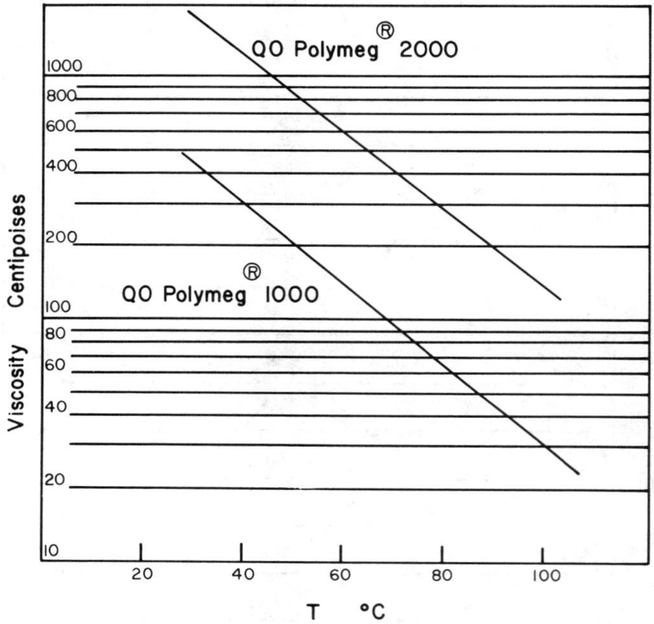

FIGURE 7.3 Viscosity against temperature for "Polymeg Polyols".[2]

very slightly soluble in aliphatic hydrocarbons, and slightly soluble in water. "Polymeg 1000" is more soluble than "Polymeg 2000". Like THF, "Polymeg Polyols" normally contain butylated hydroxy toluene (150–250 ppm) to prevent peroxide formation. They should be stored under nitrogen to prevent moisture pickup, which, if present, would participate in later reactions and diminish properties.

The properties of "Teracol" polyether glycols are similar to those of "Polymeg Polyols". Selected properties are compared in Table 7.2. Du Pont manufactures four molecular weights of "Teracol" 650, 1,000, 2,000 and 2,900. Other molecular weights can be obtained on special order.

These diprimary diols are used as starting materials for the preparation of polyurethane and polyester thermoplastic elastomers.

7.2.2 Precursors with isocyanate end groups

The diprimary diols are converted to precursors with isocyanate end groups for use in the two-step synthesis of polyurethanes. These precursors are also available commercially. Typical examples are the "Adiprenes" from du Pont[34,36] and the "Vibrathanes" from Uniroyal.[48] Each of the companies supplies several types of precursor, which differ considerably in their nominal isocyanate content, viscosity and pot life at the recommended mixing temperatures. Since the total variation in viscosity is much greater than that of the glycols, it is obvious that some of them are already partially chain-extended. The properties of these materials are compared in Table 7.4.[34,36,48] Effects on properties of the resulting vulcanizates will be considered in the next section of this chapter.

Du Pont also markets a polymer with blocked isocyanate groups that can be activated by heating. This polymer has the trade name "Adiprene BL-16".[34]

Solvents for the isocyanate terminated polymers include toluene, xylene, perchloroethylene, trichloroethylene, methyl ethyl ketone, acetone, ethyl acetate and β-ethoxyethyl acetate.

7.2.3 Polyurethanes

A truly comprehensive discussion of the properties and uses of polyurethanes derived from PTHF would require much more space than can be allotted in this brief monograph. The many variables involved in their synthesis have been described already, and the properties of the polymeric starting materials were given above. The discussion in this section will focus on examples chosen to illustrate the breadth of properties produced in polyurethanes containing PTHF just by selection of method of synthesis and monomeric starting materials. Regrettably, such a limited scope means that many significant reports and papers will be omitted or alluded to very superficially.

One-shot synthesis The ratio of reactants and the choice of reactants in these syntheses are critical. When the ratio of

TABLE 7.4
Typical Properties of Precursors with Isocyanate End Groups[32,34,46]

Trade name	Nominal NCO content, %[b]	Specific viscosity Pa·S	Specific gravity at 25°C/4°C	Working life at mixing temp,[a] min
Adiprene L-42	2.8	17[c]	1.03	7
Adiprene L-83	3.25	28[c]	1.05	5
Adiprene L-100	4.1	18[c]	1.06	10
Adiprene L-167	6.3	6.0[c]	1.07	5
Adiprene L-200	7.5	8.0[c]	1.09	3
Adiprene L-213	9.3	10[c]	1.11	1.5
Adiprene L-315	9.45	15[c]	1.11	1
Adiprene L-325	9.15	20[c]	1.11	2.5
Adiprene L-700	5.5	8.5[c]	1.07	2.5
Adiprene L-767	7.8	7.0[c]	1.09	1
Vibrathane B-600	4.0–4.3	15,[d] 6.0[e]	1.06	—
Vibrathane B-601	6.05–6.55	6.0,[d] 2.0[e]	1.07	—
Vibrathane B-602	2.95–3.15	20,[d] 8.0[e]	1.04	—

[a]Recommended with MBCA curing agent.

[b]The percentage NCO is defined as the percentage by weight of NCO groups present in the sample. One way of determining the percentage NCO is as follows. Dissolve about 1 g of the sample in 25 ml dry toluene. Add 25 ml 0.1 N di-*n*-butylamine solution in toluene and agitate for 10–15 minutes to complete the reaction. Add 100 ml isopropyl alcohol and few drops of bromophenol blue indicator (prepared from 0.1 g dry indicator, 1.5 ml 0.1 N NaOH, and diluted to 100 ml). Titrate with 0.1 N HCl solution to a yellow-green endpoint. Run a blank in the same manner. Calculate the percentage NCO $= [(B-V) \times N \times 4.202]$/weight of sample (g); B and V are the ml of 0.1 N HCl for blank and sample, respectively; N is the normality of the HCl solution.

[c]At 30°C.

[d]At 70°C.

[e]At 100°C.

PTHF glycol plus aliphatic glycol to diisocyanate is one, and the system is kept free from water and acidic materials, readily processible elastomers with good properties and no need for further vulcanization are produced. MDI and bitolylene-p,p′-diisocyanate give products with desirable properties, whereas *p*-phenylene, toluene and durene diisocyanates do not. Examples of the properties that can be developed are given in Table 7.5.[35]

Synthesis from precursors with isocyanate end groups. The properties of polyurethanes vary with the chemical structure of the starting materials and with the amount of chain-extending and curing agents. The curing temperature also affects the properties. The effect of varying the isocyanate precursor is illustrated in Table 7.6.[34] "Adiprene L" and MBCA curing agent are used as the examples. As the NCO content increases, there is a corresponding increase in hardness, 100% modulus, tensile strength, tear strength, abrasion resistance and brittleness temperature. The percentage elongation decreases.

The type of material used to cure the isocyanate precursor is an important factor in controlling the vulcanizate properties.[36,39] Three main classes are used: diamines, polyols and moisture. Of these, diamines and polyols are the most common. Differences among the curing systems are shown in Table 7.7. Specific examples will be given in Tables 7.8 and 7.9.

Aliphatic diamines react almost instantaneously and are impractical to process. Aromatic diamines react rapidly and give tough, tear- and abrasion-resistant stock. The introduction of chlorine groups into positions ortho to the amine groups reduces the reactivity of the amine groups to a desirable level. MBCA is an excellent curing agent[36] but, because of its alleged carcinogenic properties,[49] other amines are replacing it in many applications. Glycols react slowly and give softer stocks with poorer tear strength and sliding abrasion resistance. The properties of moisture cured vulcanizates resemble those of diamine cures, since ureas are formed in both cases.

TABLE 7.5
Polyurethanes Resulting from One-shot Syntheses[33]

Materials	Moles	Molecular weight	Shore A hardness	Tensile strength, MPa	Elongation, %	Modulus 300%, MPa	Wt. loss in Taber abrasion resistance apparatus,[a] g	Angle tear, kN/m	Nature
PTHF glycol	1.0	3020	65	44.1	650	6.2	~0	51	Thermoplastic
1,4-Butanediol	1.0	—							
MDI	2.0	—							
PTHF glycol	1.0	3020	—	33.8	730	—	—	58	Thermoplastic
1,10-Decanediol	1.0								
MDI	2.0								
PTHF glycol	1.00	890	83	37.9	650	7.1	—	7.1	Thermoplastic
1,4-Butanediol	1.06								
MDI	2.06								
PTHF glycol	1.0	3020	72	46.9	710	8.3	—	91	Thermoplastic
1,4-Butanediol	1.0								
Bitolylene diisocyanate	2.0								

[a]Maximum load and coarsest wheel.

TABLE 7.6
Typical Vulcanizate Properties of "Adiprene-L" Cured[32] with MBCA[a]

	Type of "Adiprene"				
	L-42	L-100	L-167	L-200	L-315
NCO content, %	2.8	4.1	6.3	7.35	9.45
MBCA, parts/100 parts glycol[b]	8.8	12.5	19.5	23.2	26.0
Hardness, durometer A	80	90	95	—	—
durometer D	—	43	48	58	73
Modulus, 100%, MPa	2.8	7.6	12.4	20.7	33.4
Tensile strength, MPa	20.7	31.0	34.5	57.2	62.0
Elongation, %	800	450	400	320	210
Tear strength,[c] kN/m	12.2	13.1	26.2	23.6	19.2
Resilience (rebound), %	—	45	40	—	45
NBS abrasion index, % of standard	110	175	300	370	285
Brittleness temperature, °C	−90	<−70	<−70	—	−64

[a]All polymers were cured at 100°C but the cure time varied.

[b]The concentration of MBCA is about 90–95% of the theoretical stoichiometric amount required to react with the isocyanate groups. Under these conditions the excess isocyanate in the system reacts with a substituted urea group to form a biuret structure.

[c]ASTM D-470.

TABLE 7.7
Comparisons of Curing Systems for "Adiprene L-100"[39]

	Diamine	Polyol	H_2O[a]
Pot life,[b] min	1–20	60–480	c
Cure time, 100°C, hours	1–3	6–24	125–250
Tensile strength, MPa	>31.0	13.8–17.2	>31.0
Hardness, Shore A	88–92	55–60	85

[a]Exposed to air at 25°C and 50% relative himidity.

[b]Time required for the polymer to become too viscous to pour, measured from the time of addition of the curing agent.

[c]Indefinite in the absence of moisture.

TABLE 7.8
Effect of Adipic Acid on MBCA-cured "Adiprene L-100"[34]

A. *Compound*				
"Adiprene L-100"	100	100	100	100
MBCA	12.5	12.5	12.5	12.5
Adipic acid	0.0	0.2	0.0	0.2
B. *Mixing and Curing*				
Mix temperature, °C	100	100	60	60
Cured, min/°C	60/100	15/100	—	—
Cured, days/°C	—	—	1/24, 7/24	1/24, 7/24
C. *Physical Properties*				
Hardness, durometer A	88	86	69, 80	80, 85
300% modulus, MPa	8.1	10.5	4.3, 7.2	6.6, 9.6
Tensile strength, MPa	15.2	22.4	4.6, 12.8	21.0, 30.2
Elongation at break, %	800	500	380, 560	530, 440
Compression set, %				
22 hours at 70°C	22	30	— —	— —
After cured: one week at 24°C; 50% RH:				
Hardness, durometer A	90	91	— —	— —
300% modulus, MPa	14.8	16.2	— —	— —
Tensile strength, MPa	31.7	28.0	— —	— —
Elongation at break, %	460	380	— —	— —
Compression set, %				
22 hours at 70°C			— —	— —

TABLE 7.9
Curing "Adiprene L-100" with Polyols[34]

A. *Compounds*						
"Adiprene L-100"	100	100	100	100	100	100
1,4-Butanediol	4.0	3.0	3.5	3.5	3.5	3.5
Trimethylolpropane	0.3	1.3	0.8	0.8	0.8	0.8
Ferric acetylacetonate	—	—	—	0.1	—	—
Triethylenediamine	—	—	—	—	0.01	—
Stannous octoate	—	—	—	—	—	0.01
B. *Mixing and Curing*						
Mixed at 100°C						
Cured: hrs/°C	4/100	4/100	16/100	6/100	6/100	6/100
Postcured: hrs/°C	16/100	16/100	—	—	—	—
Aftercured, days[a]	7	7	7	7	7	7
Pot life at 100°C, min	45	30	120	8	50	19
Demolding time at 100°C, min	90	90	240	20	80	30
C. *Physical Properties*						
Hardness, durometer A	57	60	60	58	58	58
Modulus, 100%, MPa	1.4	1.4	1.9	1.7	1.9	1.6
Tensile strength, MPa	8.9	10.3	14.5	13.8	13.4	11.2
Elongation at break, %	560	470	500	450	500	560
Tear strength,[b] kN/m	—	—	2.1	2.6	2.1	2.4
Compression set, %	18	6	9	13	8	9

[a]Stored at 24°C and 50% RH.
[b]ASTM-470.

Aromatic diamines are the most common curing agents and, of these, MBCA is the most widely used.[36] Polymers cured in this way contain urea groups along the chain, which are potentially able to form hydrogen bonds between molecular chains. These physical "crosslinks" are responsible for the high strength of diamine cured resins. Chemical crosslinks in the form of biuret linkages also contribute to strength, if they are not too numerous. Biuret linkages will be present whenever the number of amine groups is less than the total number of isocyanate groups.

The effect of MBCA concentration on the physical properties of "Adiprene L-100" vulcanizates is shown in Figure 7.4. The best balance of properties is obtained with 90–95% of the theoretical amount of MBCA.

Addition of a small amount of a carboxylic acid such as adipic acid to an MBCA recipe accelerates the curing reaction and permits a reduction in demolding time. (Demolding time is the cure time after which a molded part can be removed from a mold without permanent distortion.) Lower curing temperatures also become possible.

Shorter cure times have production advantages; but a shorter cure time also means a shorter pot life. Some effects of the addition of adipic acid to MBCA recipes are indicated in Table 7.8, which also shows that molded parts should be stored for at least a week before putting into service, since ultimate properties may not be attained until then.

Diamines are effective curing agents in the presence of plasticizers such as dioctyl phthalate or triethylene glycol-di(2-ethyl hexoate). Mixing and curing can be effected at moderately low temperatures by adding adipic acid or using methylene dianiline, which is more reactive than MBCA.[36]

Binary and ternary mixtures of diamines have also been used to cure "Adiprene L-100".[36] Such mixtures can usually be handled at lower temperatures than are possible with either diamine alone. The processing and vulcanizate properties lie between those obtained with the individual diamines. Ternary-amine blends typically give softer vulcanizates than are obtained with a single diamine or with binary mixtures. This is

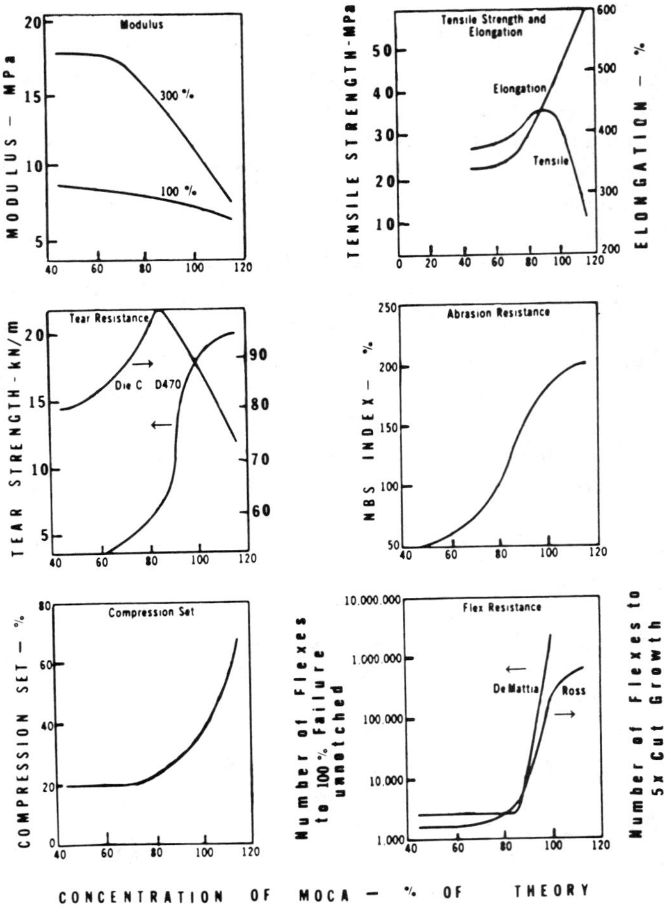

FIGURE 7.4 Effect of concentration of MBCA on physical properties of "Adiprene L-100" vulcanizates.[36] (MOCA is du Pont's trade name for MBCA. It is no longer produced.)

probably caused by an increase in chain irregularity, which decreases intermolecular bonding in much the same manner as too many biuret linkages.

Polyol curing systems for diisocyanate terminated precursors consist of mixtures of diols in combination with triols. The triols provide the branch sites that occur spontaneously in the form of biuret crosslinks with diamines. Branching due to similar reactions with diols is insignificant. As already noted in relation to Table 7.7, the vulcanizates tend to be softer than those obtained after diamine curing. Polyol curing is slower than diamine curing, and results in longer pot life and longer demolding times.[36] The curing rates can be accelerated by adding materials such as triethylenediamine, stannous octoate or ferric acetylacetonate. Some typical polyol recipes and the properties that result are given in Table 7.9. Because of their greater initial softness, polyol cured polyurethanes retain their flexibility at lower temperatures (−73°C) better than diamine-cured polyurethanes. Compared to polyurethanes from polyesters, the polyether based polyurethanes containing PTHF have advantages in resistance to degradation by moisture, processing ease, resilience, toughness and abrasion resistance.[36,50]

Many of the uses of polyurethanes derived from PTHF glycols were already mentioned in the introduction to this section. Since properties can be varied widely by choice of method of synthesis and curing, the polyurethane can be tailor-made for a given application. These polyurethane properties decrease at elevated temperature, but they are good (up to about 107°C) compared to conventional elastomers. At higher temperatures, properties deteriorate rapidly. Resistance to weather, fungus, gamma radiation and aliphatic hydrocarbon are excellent, but deterioration occurs in direct sunlight unless a protecting agent such as carbon black is added. Aromatic and polar solvents cause severe swelling. Electrical properties are satisfactory for use in potting and encapsulating applications.[36] The Quaker Oats Company has compiled a partial list of references and abstracts of those references on the use of PTHF glycols in formulating polyurethanes.[51] Uses in elastic fibers (e.g., "Spandex" from du Pont), elastomers, flexible foams, adhesives and coatings are suggested. Such a detailed discussion does not seem appropriate here, but their list of references has been included.[50,52–95]

Any discussion of the properties of polyurethanes would be

incomplete without some mention of *morphology*. Earlier we stated that polyurethanes are an example of thermoplastic elastomers composed of segmented molecules in which soft and hard segments are randomly distributed along the backbone. The soft segments in the polyurethanes under discussion in this chapter are the PTHF. The hard segments are the products of reaction of the aromatic diisocyanate and the low molecular weight glycol or diamine. The idea that the hard segments are held together by hydrogen bonds or "virtual crosslinking" was suggested by Schollenberger and coworkers[96] in 1958. However, it was not until about 1966 that Cooper and Tobolsky,[91] Flocke[98] and Shimanski and coworkers[99] reported experimental evidence that polyurethanes do indeed consist of a two phase structure consisting of high modulus domains in a rubbery matrix, and that the domains were held together by secondary valence forces. Since then, extensive work by Bonart and coworkers,[100–3] and many others[104–11] has led to a much greater understanding of the physical structure of these polyurethanes and of how they behave on stretching. Evidence has been gathered from x-ray diffraction studies, DSC (differential scanning calorimetry) measurements, infrared dichroism studies, ultra-low frequency AC measurements, dynamic mechanical responses and microscopy.

The structural model proposed by Bonart and coworkers[100–3] to explain their observations is shown in Figure 7.5. It consists of soft segments with hard segment domains in layers, plus areas with hard segments dissolved in the soft segment matrix and areas where soft segment chains run through the hard segment layers. The arrangement of the atoms within the hard segment layers for the hydrogen bonding in MDI–butanediol segments is shown in two dimensions in Figure 7.6 and in three dimensions in Figure 7.7. Recently, Blackwell and Gardner[105] have added a further refinement to the structure of the hard segment layer by proposing the packing of the chains shown in Figure 7.8. Successive monomer units are related by a center of symmetry at the central CH_2–CH_2 bond. The chains are linked in stacks through C=O · · · H–N hydrogen bonds which involve half the urethane groups, A. The remaining

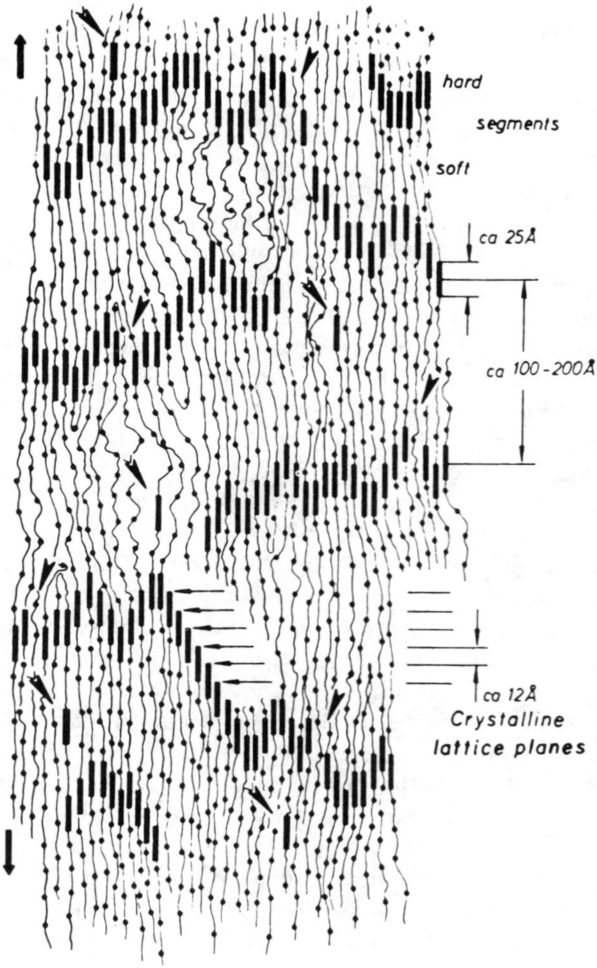

FIGURE 7.5 Structural model for reciprocal arrangement of the hard and soft segments in elongated polyurethane elastomers. Key: (➘) hard segments dissolved in the soft-segment matrix. (➚) soft-segment chains which run through the hard-segment layers and contribute to their structure.[102] (Reprinted from R. Bonart, L. Morbitzer and E. H. Mühler, J. Macromol. Sci.-Phys., *B9*, 447 (1974) by courtesy of Marcel Dekker, Inc.)

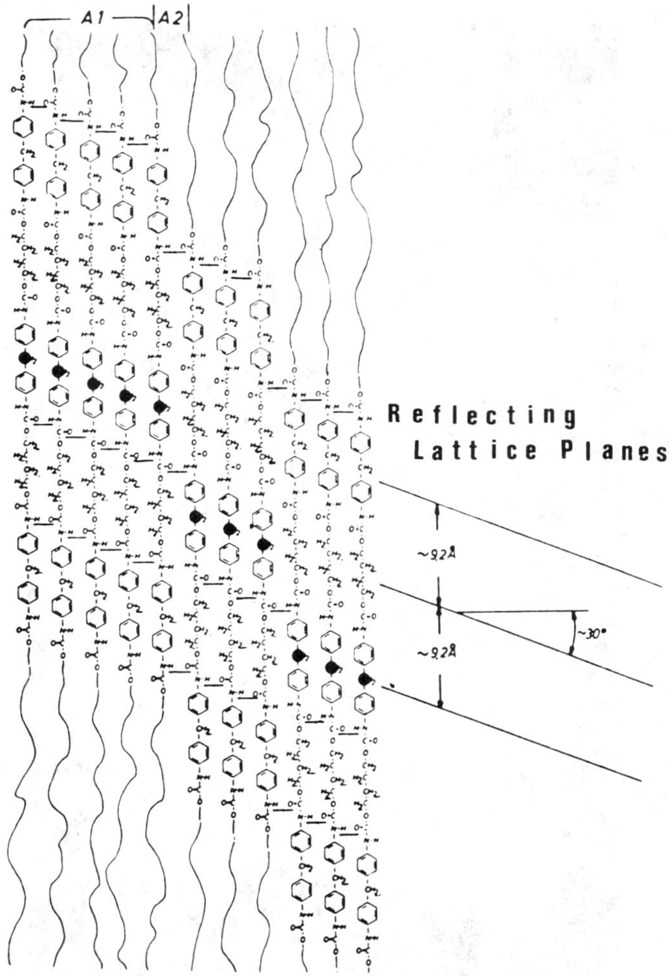

FIGURE 7.6 Schematic drawing of the hydrogen-bonding in MDI–butanediol hard segments, as proposed by Bonart *et al.*[101] The staggering of chains leads to planes in this projection at 30° to the perpendicular as indicated. (Reprinted from R. Bonart, L. Morbitzer and G. Hentze, J. Macromol. Sci.-Phys., *B3*, 337 (1969) by Courtesy of Marcel Dekker, Inc.)

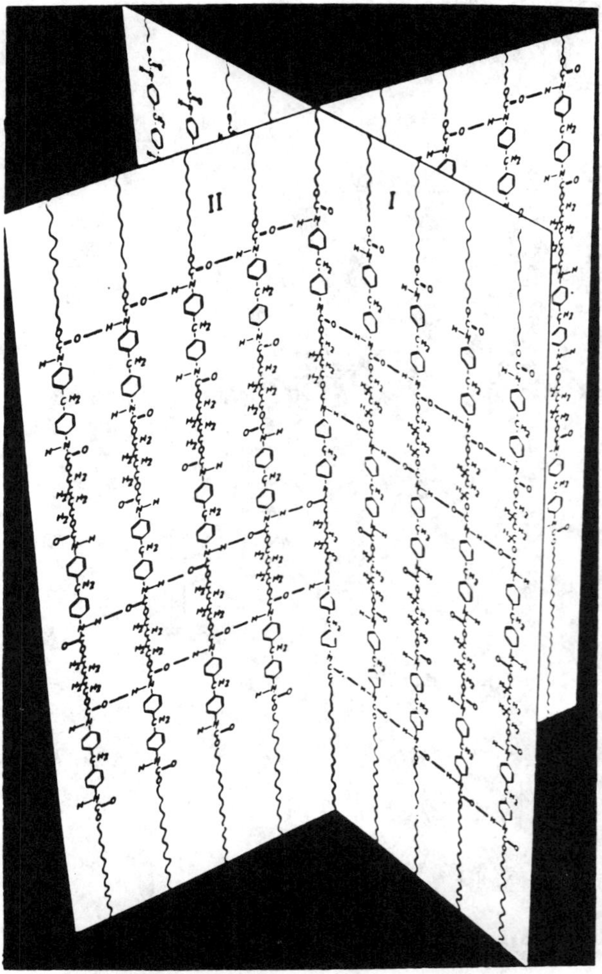

FIGURE 7.7 Diagrammatical representation of the three-dimensional cross-linking structure of a polyurethane extended with 1,4-butanediol.[102] (Reprinted from R. Bonart, L. Morbitzer and E. H. Mühler, J. Macromol. Sci.-Phys., *B9*, 447 (1974) by courtesy of Marcel Dekker, Inc.)

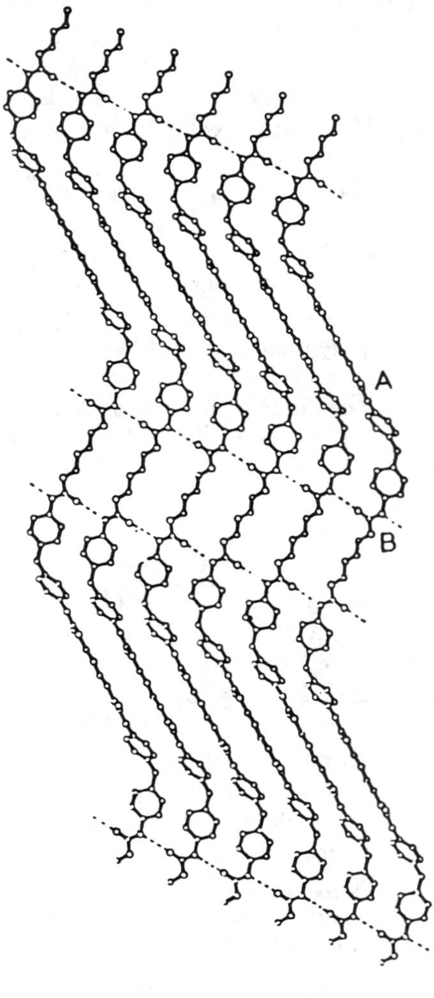

FIGURE 7.8 Packing of chains viewed along the *a* axis of the unit cell.[105] (Reproduced from J. Blackwell and K. H. Gardner, *Polymer, 20,* 13 (1979) by permission of the publishers, IPC Business Press Ltd. ©)

urethanes, B, are similarly hydrogen-bonded to adjacent stacks. Thus, the structure is stabilized by hydrogen-bonding in both directions perpendicular to the chain direction. A triclinic unit cell is proposed for the structure with dimensions $a = 5.2$ Å, $b = 4.8$ Å, $c = 35.0$ Å, $\alpha = 115°$, $\beta = 121°$ and $\gamma = 85°$. The space group is $P\bar{1}$ and the cell contains two repeat units of a single chain.[105]

In general, polyurethanes show high hysteresis. The morphology above helps to explain these results. On stretching, first soft segments stress crystallize. On further stressing, the hard segments break up and reform in a manner which relieves stress.[104] Once the stress is relieved, the soft segments recover well, but the hard segments do not. Differences in the hysteresis behavior of polymers with variable amounts of isocyanate and effects of different molecular weight PTHF glycols on properties can also be explained on the basis of morphology changes.[106,112]

7.2.4 Polyesters

Thermoplastic elastomers based on PTHF glycols and other polyalkylene ether glycols are relatively new materials. Development work and sustained interest in them started at the end of the 1960s, and has continued ever since.[5,41–45,113–21] One polymer based on 55% soft segment from PTHF glycol of about 1,000 molecular weight and 45% hard segment derived from 1,4-butanediol and methylterephthalate is finding increased use in commercial application under the du Pont trade name "Hytrel". Toboyo Co. in Japan produces a similar polymer called "Pelprene".[6] A more recent entry into the commercial market is Eastman Kodak's "EKTAR", which has 1,4-cyclohexanedimethanol in its hard segment.

The studies of these copolymers have included an examination of the effect of varying the molecular weight of the PTHF glycol, comparisons with polyesters derived from other polyalkylene glycols, and studies of the effect of chemical structure of diol and/or diacid on ease of synthesis and properties. Morphology and other physical properties have also been examined.

The molecular weight of PTHF glycol used in the synthesis of these block copolymers has an important influence on copolymer properties.[43] A glycol of about 1,000 molecular weight leads to copolymers with the best overall properties, especially resistance to stiffening. When PTHF glycols of 3,000 molecular weight or more are used, the copolymers contain crystalline polyether regions at room temperature. At approximately 2,000 molecular weight, stress-induced crystallization occurs on stretching to 100% or more at room temperature. Copolymers based on PTHF glycols of very low molecular weights also have higher stiffening temperatures and higher moduli than would be expected from the ratio of ingredients charged. These increases occur because some of the glycol is lost by volatilization during the final stages of the polymerization and the final content of, for example, tetramethylene terephthalate units is higher than expected. Some trends with glycol molecular weight are illustrated in Figure 7.9 for copolymers containing 57 wt.% of tetramethylene terephthalate units plus 43 wt.% PTHF terephthalate units.[43]

Table 7.10A shows a comparison of properties of copolymers containing 57 wt.% of poly(tetramethyl terephthalate) plus

TABLE 7.10
Copolymer Properties as a Function of Polyether Structure[43]

Property	Polyether glycol		
	PTHF	Polyethylene oxide	Polypropylene oxide
A. For 43% glycol terephthalate:			
Stress at 100%, MPa	13.4	13.2	11.7
Tensile strength, MPa	47.2	42.6	27.2
Elongation, %	660	705	640
Permanent set, %	365	190	355
Tear strength, kN/m	62	31	18
B. For 26% glycol terephthalate:			
Stress at 100%, MPa	20.7	21.8	21.2
Tensile strength, MPa	37.6	50.0	39.3
Elongation, %	380	515	456
Permanent set, %	284	303	315
Tear strength, kN/m	87	62	42

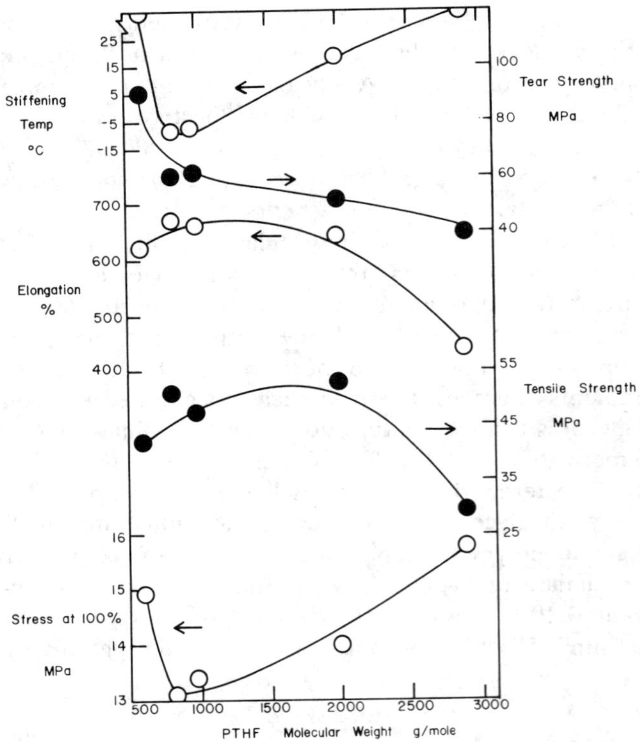

FIGURE 7.9 Properties of 57% tetramethylene terephthalate copolymers prepared from PTHF glycols of different molecular weight. Data from Ref. 43.

43 wt.% terephthalate from different polyalkylene glycols.[43] The molecular weights of the starting glycols were all about 1,000. The properties of copolymers from PTHF, especially tear strength and tensile strength at break, are outstanding. The properties of copolymers from polypropylene glycol are low because this polyether degrades relatively rapidly at the temperatures used for the final steps in its preparation.[122] The copolymers from PTHF also have greater hydrolytic stability because of the lower oxygen-to-carbon ratio.

When the concentration of the poly(tetramethyl terephthalate) is increased to 74%, the differences in properties due to changes in structure of the polyglycol are smaller.[43] The data are given in Table 7.10B. The tear strength of copolymers containing PTHF remains relatively high.

In a more extensive study of polymers based on PTHF glycol of 1,000 molecular weight and covering a concentration range from 70 to 18% PTHF terephthalate, effects of concentration of poly(tetramethylene terephthalate) were better defined.[44] Stress at 100% elongation decreased with increasing concentration of the PTHF glycol units. At 43.7 wt.% glycol ester and below, the copolymers showed yield points below 100% elongation. The yield points may be attributed to changes in the orientation of the crystalline hard phase segments. Tear strength, hardness and melting point of the hard segment phase increased with decreasing content of the polyglycol in the copolymers. The compression set was highest at 70% polyglycol content and changed very little below 60% polyglycol.[44]

The copolymers stiffened more readily as the concentration of polyglycol decreased and the glass transition temperature of the amorphous phase of the copolymers increased. The cause of the increase has been attributed to higher concentrations of uncrystallized poly(tetramethylene terephthalate) in the amorphous phase,[119–21] the increased number of crystalline tie points, and the greater reinforcement of the amorphous phase by the longer poly(tetramethylene terephthalate) segments. Approximately 35–55% of the total poly(tetramethylene terephthalate) segments appear to crystallize.[44] The higher percentages naturally are found in samples containing larger weight percentages of the latter polyester, which as a homopolymer is a tough, rapidly crystallizing plastic.

Elongation at break decreases and the block copolymers become less elastic as the polyglycol concentration in the copolymer decreases.

The structure of the monomeric glycol has an important effect on the properties of the block copolymers.[44] Copolymers based on 1,4-butanediol excel in the ease of synthesis, rapid rate of crystallization from the melt, high melting point, excellent tensile strength and excellent tear strength. Some comparisons with properties of block copolymers prepared using other monomeric glycols are given in Table 7.11. The copolymer from ethylene glycol is similar to that from 1,4-butanediol, except that it crystallizes slowly. The problem with a slow rate of crystallization is that it limits the use of injection-molding techniques in the fabrication of molded goods.[44] Other aliphatic α,ω-diols gave copolymers which were relatively low in tensile and tear strengths. Copolymers from 1,4-benzenedimethanol and 1,4-cyclohexanedimethanol were relatively lower in inherent viscosity, tensile strength and tear strength. Copolymers from 1,4-cyclohexanedimethanol were brittle at the content of diol where those from other monomeric glycols had good properties.

Not all block copolyesters were prepared from terephthalate esters. Other aromatic esters such as isophthalates, 2,6-naphthalenedicarboxylates, 4,4′-biphenyldicarboylates and *m*-terphenyl-4,4″-dicarboxylates have been used.[44] The structure of the ester segment had an important effect on ease of synthesis and ultimate molecular weight. This can be seen by comparing the data for isophthalates given in Table 7.12 with those for the terephthalates in Table 7.11. Properties with isophthalate copolymers containing 1,4-butanediol are still very good, but the melting temperature is significantly reduced compared to the terephthalates. But now most properties of isophthalate copolymers containing 1,4-cyclohexanedimethyl esters are equal to, or better than, those from copolymers containing esters from 1,4-butanediol. Only the tensile strength is significantly lower. The reason for the difference is that, with 1,4-cyclohexanedimethanol, phase separation occurs in polymerizing melts of terephthalate-based copolymers, but not in those of isophthalate-based copolymers. The principal disadvantage of these copolymers is that they are slow to crystallize.[44]

As with the polyurethanes a variety of experimental techniques have been applied to the study of the morphology of polyester thermoplastic elastomers. Efforts have been made to

TABLE 7.11

Properties of Polyester Block Copolymers as a Function[42] of Diol Structure[a]

Diol	Inherent viscosity, l/g	Compression set, %	Stress at 100%, MPa	Tensile strength, MPa	Elongation, %	Permanent set, %	Tear strength, kN/m	Melting T, °C	Stiffening T, °C
$HO(CH_2)_2OH$	0.13	52	11.4	45.5	675	275	42	224	−38
$HO(CH_2)_3OH$	0.17	48	11.7	22.8	660	195	15	198	−36
$HO(CH_2)_4OH$	0.18	52	11.7	48.4	755	370	48	189	−33
$HO(CH_2)_5OH$	0.16	90	4.5	15.4	880	210	25	106	−50
$HO(CH_2)_6OH$	0.15	86	5.2	13.5	750	370	18	122	−53
$HO(CH_2)_{10}OH$	0.13	81	6.1	15.5	640	370	7.8	106	−51
$HOCH_2C_6H_{10}CH_2OH$[b,c]	0.12	62	6.2	11.0	470	119	6.0	—	−60
$HOCH_2C_6H_4CH_2OH$[b]	0.10	58	11.7	21.3	380	143	14.4	227	−42

[a]All copolymers except the one from 1,4-cyclohexanedimethanol had 50% alkylene terphthalate and were prepared from a PTHF glycol of about 1,000 molecular weight. Data for the copolymer from 1,4-cyclohexanedimethanol is for 30% alkylene terephthalate. Above that concentration, the copolymers were brittle and had very low elongation.

[b]1,4-derivatives.

[c]Practical grade, which has about a 70/30 mixture of *trans* to *cis* monomer. Most *trans* copolymers were brittle.

TABLE 7.12
Properties of 50% Alkylene Isophthalate Copolymers[42]

Properties	Diol		
	$HO(CH_2)_4OH$	$HOCH_2C_6H_{10}CH_2OH$[a,b]	$HOCH_2C_6H_{10}CH_2OH$[a,c]
Inherent viscosity, l/g	0.16	0.14	0.11
Stress at 100%, MPa	7.2	5.7	6.9
Tensile strength, MPa	58.6	31.7	33.8
Elongation, %	720	690	760
Permanent set, %	126	70	179
Tear strength, kN/m	54	78	61
Compression set, %	91	89	87
Melting temperature, °C	85, 112	147	184
Stiffening temperature, °C	−31	−38	−28

[a]1,4-derivatives.
[b]Practical.
[c]100% *trans*.

relate the observed morphology to chemical structure and physical properties.[116–7,119]

The studies on solution and melt-crystallized samples have shown that the materials have a spherulitic-lamellar texture with a two-phase structure consisting of continuous and interpenetrating crystalline and amorphous domains. The lamellar domains formed by crystallization of the tetramethylene terephthalate hard segments are approximately 100 Å in width and up to several thousand angstroms in length.[119] There is considerably more interpenetration between the hard and soft segments of polyester thermoplastic elastomers than there is in polystyrene–polybutadiene–polystyrene triblock polymers. The crystalline regions are continuous and highly interconnected, and serve as intermolecular tie points for the amorphous portions of the polymer. At temperatures below the melting temperature of the polyester segments, these polyesters behave much like conventionally cured elastomers. Above the melting temperature of the polyester segments, they flow and can be processed like a thermoplastic.[119]

The spherulitic structure is disrupted on stretching. Ghaffar *et al.*[117] have used the "string of beads" model shown in Figure 7.10 to explain the properties and morphology of polyester thermoplastic elastomers. Crystalline concentrations

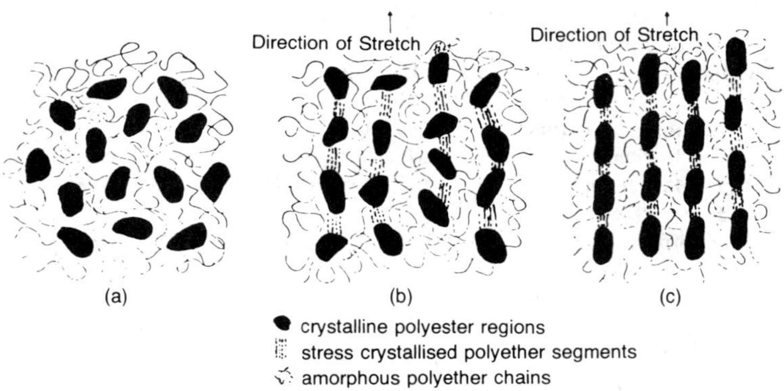

FIGURE 7.10 Schematic model of a polyester thermoplastic elastomer: (a) unstretched; (b) at 200% extension; (c) at high extensions.[117]

of polyester, the "beads", lie in all possible orientation within an amorphous magma composed principally, but not exclusively, of polyether. At low extensions (~200%) the amorphous polyether segments stress-crystallize with their chains parallel to the stress direction, but no orientation of the polyester domains occurs. On further stretching the polyester "beads" align into nearly parallel "strings of beads".

REFERENCES

1. "Polymeg Polyols for High-performance Urethane Elastomers", Form No. 10131/8/1976, Chemicals Division, The Quaker Oats Co., Chicago, IL 60654 (1976).
2. "QO Polymeg", Bulletin 207, Chemicals Division, The Quaker Oaks Co., Chicago, IL 60654, (undated).
3. "QO Polymeg", Technical Bulletin No. 150-B, Chemicals Division, The Quaker Oats Co., Chicago, IL 60654 (1967).
4. "Teracol Polyether Glycol", Product Data Sheet, Industrial Chemicals Department, E. I. du Pont de Nemours and Co., Wilmington, Del. 19898, (undated).
5. *du Pont Innovation*, **4**(1), 13 (1972).
6. K. Matsuda, *Chemtech*, **4**, 744 (Dec. 1974).
7. K. Matsuda, Y. Tanaka and T. Sakai, *J. Appl. Polym. Sci.*, **20**, 2821 (1976).
8. H. Meerwein, German Patent 766,208 (June 29, 1953); through *Chemisches Zentralblatt*, **131**, 5993 (1960).
9. G. Pruckmayr and T. K. Wu, *Macromolecules*, **11**, 265,662 (1978).
10. J. Guzman, M. A. Sanchez and J. G. Fatou, *An. Quim.*, **74**(9–10), 1155 (1978); through *CA*, **91**, 5545v (1979).
11. F. B. Hill and R. A. Schulze, US Patent 2,751,419 (to E. I. du Pont de Nemours and Co.) (June 19, 1956).
12. T. Shono, T. Tsujino and Y. Hachihama, *Kogyo Kagaku Zasshi*, **61**, 1347 (1958); cited in Y. Tanaka, *J. Macromol. Sci.-Chem.*, **A11**, 2189 (1977) or Y. Hachihama and T. Shono, *Technology Reports of the Osaka University*, **9**(361), 28 (1959).
13. W. Stix and W. Heitz, *Makromol. Chem.*, **180**, 1367 (1979).
14. W. Öchsner and K.-H. W. Reichert, *Makromol. Chem.*, **150**, 1 (1971).
15. P. Dreyfuss, *Chemtech*, **3**, 356 (June, 1973).
16. P. Dreyfuss and M. P. Dreyfuss, US Patent 3,548,037 (to B. F. Goodrich Co.) (Dec. 15, 1970).
17. E. I. du Pont de Nemours and Co., Belgian Patent 868,726 (Jan. 4, 1979).
18. E. I. du Pont de Nemours and Co., Belgian Patent 868,381 (Dec. 22, 1978).

19. G. Pruckmayr, US Patent 4,153,786 (to E. I. du Pont de Nemours and Co.) (May 8, 1979).
20. G. Pruckmayr and R. H. Weir, US Patent 4,120,903 (to E. I. du Pont de Nemours and Co.) (Oct. 17, 1978).
21. Daniel J. Vaughan, " 'Nafion', an Electrochemical Traffic Controller", E. I. du Pont de Nemours and Co., Wilmington, Del. 19898, (undated).
22. G. E. Heinsohn, I. M. Robinson, G. Pruckmayr and W. H. Gilbert, US Patent 4,163,115 (to E. I. du Pont de Nemours) (July 31, 1979).
23. Y. Yamashita, M. Hirota, K. Nobutoki, Y. Nakamura, A. Hirao, S. Kozawa, K. Chiba, H. Matsui and G. Hattori, *J. Polym. Sci.*, **B8**, 481 (1970).
24. Y. Yamashita, M. Hirota, H. Matsui, A. Hirao and K. Nobutoki, *Polym. J.*, **2**, 43 (1971).
25. S. Smith and A. J. Hubin, *J. Macromol. Sci.-Chem.*, **A7**, 1399 (1973).
26. R. Hoene and K.-H. W. Reichert, *Makromol. Chem.*, **177**, 3545 (1976).
27. E. Franta, L. Reibel, J. Lehmann and S. Penczek, *J. Polym. Sci.: Polym. Symp.*, **56**, 139 (1976).
28. J. Lehmann, L. Reibel, E. Franta, W. Dobrogoszcz and S. Penczek, "Abstracts to First International Symposium on Polymerization of Heterocycles (Ring-opening)", Warsaw-Jablonna, Poland, p. 141 (June 23–25, 1975).
29. D. H. Richards, S. B. Kingston and T. Souel, *Polymer*, **19**, 68 (1978).
30. P. Dreyfuss and J. P. Kennedy, *J. Polym. Sci.: Polym. Lett. Ed.*, **14**, 139 (1976).
31. M. Kučera, F. Bozek and K. Majerová, *Polymer*, **20**, 1013 (1979).
32. M. Kučera and H. Kelblerová, *Coll. Czech. Chem. Commun.* (in press).
33. QO Polymeg, Bulletin 208, "Liquid TDI Castable Elastomers", Chemicals Division, The Quaker Oats Co., Chicago, IL 60654, (undated).
34. "Types of 'Adiprene' L and 'Adiprene' LW", AP-210.1 (R1), Elastomers Chemicals Department, E. I. du Pont de Nemours and Co., Wilmington, Del. 19898, (undated).
35. C. S. Schollenberger, US Patent 2,899,411 (to B. F. Goodrich Co.), (Aug. 11, 1959).
36. "Adiprene L-100", AP-220.L100, Elastomers Chemicals Department, E. I. du Pont de Nemours and Co., Wilmington, Del. 19898 (1976).
37. P. Wright and A. P. C. Cumming, "Solid Urethane Elastomers", MacClaren and Sons, London (1969).
38. K. C. Frisch and S. L. Reegan (eds.), "Advances in Urethane Science and Technology", Technomic Publishing Co., Stamford, Conn. 06901, (1971).
39. R. J. Athey, Contribution No. 138, Division of Rubber Chemistry, American Chemical Society, Cincinnati, OH (1958).
40. R. J. Athey, *Ind. Eng. Chem.*, **52**, 611 (1960).
41. M. Brown and W. K. Witsiepe, *Rubber Age*, p. 35 (Mar. 1972).
42. W. K. Witsiepe, *Adv. Chem. Series*, **129**, 39 (1973).

43. J. R. Wolfe, Jr., *Rubb. Chem. Technol.*, **50**, 688 (1977).
44. J. R. Wolfe, Jr., *Adv. Chem. Series*, **176**, 129 (1979).
45. Eastman Kodak Co., Neth. Appl. 6,508,295 (Cl.C 08 g) (Dec. 30, 1965).
46. FDA Regulation 121.2550.
47. FDA Regulation 121.2562.
48. "Vibrathane B-600, B-601 and B-602, Polyether-based Urethane Liquid Polymers", Form No. 740-B5, Chemical Division, US Rubber Co., Naugatuck, Conn. 06771, (undated).
49. C. E. Searle (ed.), "Chemical Carcinogens", ACS Monograph 173, Am. Chem. Soc., Washington, DC (1976).
50. R. J. Ferrari, *Rubber Age*, p. 53 (Feb. 1967).
51. "A Partial List of References on Polytetramethylene Ether Glycols", 150-B, Addendum No. 2, The Quaker Oats Co., Chicago, IL 60654, (undated).
52. S. L. Axelrood, L. C. Smith and K. C. Frisch, *Rubber Age*, p. 233 (Nov. 1964).
53. C. M. Barringer, US Patent 3,248,373 (to E. I. du Pont de Nemours and Co.) (Apr. 26, 1966).
54. B. D. Beitchman and W. R. Erner, US Patent 3,192,187 (to Air Products and Chemicals, Inc.) (June 29, 1965).
55. A. Bell, C. J. Kibler and J. G. Smith, US Patent 3,243,413 (to Eastman Kodak Co.) (Mar. 29, 1966).
56. A. Bell, C. J. Kibler and J. G. Smith, US Patent 3,261,812 (to Eastman Kodak Co.) (July 19, 1966).
57. R. L. Berger and M. A. Youker, US Patent 3,178,310 (to E. I. du Pont de Nemours and Co.) (Apr. 13, 1965).
58. D. Bianca and R. E. Knox, *Rubber Age*, p. 76 (May, 1966).
59. J. G. Burt and H. C. Walter, US Patent 3,287,330 (to E. I. du Pont de Nemours and Co.) (Nov. 22, 1966).
60. C. W. Croco, US Patent 2,901,467 (to E. I. du Pont de Nemours and Co.) (Aug. 25, 1959).
61. E. Csendes, C. Hundred and J. J. Verbanc, US Patent 3,021,301 (to E. I. du Pont de Nemours and Co.) (Feb. 13, 1962).
62. P. A. DePaolo, *Modern Plastics*, **43**(7), 153 (Mar., 1966).
63. R. J. Ferrari, US Patent 3,255,129 (to US Rubber Co.) (June 7, 1966).
64. A. F. Finelli, US Patent 3,085,991 (to Goodyear Tire and Rubber Co.) (Apr. 16, 1963).
65. P. E. Frankenberg and A. H. Frazer, US Patent 2, 957, 852 (to E. I. du Pont de Nemours and Co.) (Oct. 25, 1960).
66. A. H. Frazer, US Patent 3,009,901 (to E. I. du Pont de Nemours and Co.) (Nov. 21, 1961).
67. A. H. Frazer, US Patent 3,071,557 (to E. I. du Pont de Nemours and Co.) (Jan. 1, 1963).
68. H. Z. Friedlander and R. N. Rickles, *Chem. Eng.*, p. 145 (June 6, 1966).

69. G. F. Gmitter, US Patent 3,078,239 (to General Tire and Rubber Co.) (Feb. 19, 1963).
70. F. B. Hill, Jr., US Patent 2,929,800 (to E. I. du Pont de Nemours and Co.) (Mar. 22, 1966).
71. J. Hochberg, US Patent 3,257,261 (to E. I. du Pont de Nemours and Co.) (June 21, 1966).
72. O. Keplinger, O. C. Elmer and J. S. Duncan, US Patent 3,062,772 (to General Tire and Rubber Co.) (Nov. 6, 1962).
73. C. J. Kibler, A. Bell and J. G. Smith, US Patent 3,238,178 (to Eastman Kodak Co.) (Mar. 1, 1966).
74. W. Kirk, Jr., *Am. Dyestuff Reporter*, p. 59 (Sept. 16, 1963).
75. C. R. Koller, US Patent 2,813,776 (to E. I. du Pont de Nemours and Co.) (Nov. 19, 1957).
76. C. R. Koller, US Patent 2,929,801 (to E. I. du Pont de Nemours and Co.) (Mar. 22, 1960).
77. A. Lebovits, *Modern Plastics*, **43**(7), 139 (Mar., 1966).
78. M. D. Livingood, US Patent 2,850,467 (to E. I. du Pont de Nemours and Co.) (Sept. 2, 1958).
79. G. Loew, US Patent 3,222,300 (to Farbenfabriken Bayer AG and Mobay Chemical Co.) (Dec. 7, 1965).
80. D. B. Pattison, US Patent 2,830,038 (to E. I. du Pont de Nemours and Co.) (Apr. 8, 1958).
81. T. Ramos, US Patent 3,274,134 (Sept. 20, 1966).
82. K. U. Rausch, Jr. and A. A. R. Sayigh, *Ind. Eng. Chem. Prod. Res. Develop.*, **4**, 92 (1965).
83. J. E. Romano, US Patent 3,111,368 (to E. I. du Pont de Nemours and Co.) (Nov. 19, 1963).
84. J. H. Saunders and K. C. Frisch, "Polyurethanes", Vol. XVI, Part II, "Technology", Interscience, New York, NY (1964), p. 694.
85. J. C. Shivers, Jr., US Patent 3,023,192 (to E. I. du Pont de Nemours and Co.) (Feb. 27, 1962).
86. R. J. Sievers, US Patent 2,983,693 (to E. I. du Pont de Nemours and Co.) (May 9, 1961).
87. W. Steuber, US Patent 2,813,775 (to E. I. du Pont de Nemours and Co.) (Nov. 19, 1957).
88. W. Steuber, US Patent 2,929,804 (to E. I. du Pont de Nemours and Co.) (Mar. 22, 1960).
89. R. J. Thurmaier, US Patent 3,149,998 (to E. I. du Pont de Nemours and Co.) (Sept. 22, 1964).
90. J. J. Verbanc, US Patent 3,197,440 (to E. I. du Pont de Nemours and Co.) (July 27, 1965).
91. J. W. Watson and P. Ford, US Patent 3,130,102 (to Dunlop Co. Ltd.) (Apr. 21, 1964).
92. M. A. Weinberg, US Patent 3,271,352 (to Johnson and Johnson) (Sept. 6, 1966).

93. E. Windemuth and F. Moller, US Patent 3,239,480 (to Farbenfabriken Bayer AG) (Mar. 8, 1966).
94. C. A. Young, US Patent 2,858,287 (to E. I. du Pont de Nemours and Co.) (Oct. 28, 1958).
95. E. L. Yuan, US Patent 3,190,765 (to E. I. du Pont de Nemours and Co.) (June 22, 1965).
96. C. S. Schollenberger, H. Scott and S. R. Moore, *Rubber World*, p. 549 (Jan., 1958).
97. S. L. Cooper and A. V. Tobolsky, *J. Appl. Polym. Sci.*, **10**, 1837 (1966).
98. H. A. Flocke, *Kunststoffe*, **56**(5), 328 (1966).
99. V. M. Shimannskii, S. I. Shkol'nik and S. B. Kozakov, *Soviet Rubber Technol.* (English translation), **26**, 20 (1967).
100. R. Bonart, *J. Macromol. Sci.-Phys.*, **B2**, 115 (1968).
101. R. Bonart, L. Morbitzer and G. Hentze, *J. Macromol. Sci.-Phys.*, **B3**, 337 (1969).
102. R. Bonart, L. Morbitzer and E. H. Müller, *J. Macromol. Sci.-Phys.*, **B9**, 447 (1974).
103. R. Bonart, *Angew. Makromol. Chem.*, **58/59**, 259 (1977).
104. C. W. Wilkes and C. Yusek, *J. Macromol. Sci.-Phys.*, **B7**, 157 (1973).
105. J. Blackwell and K. H. Gardner, *Polymer*, **20**, 13 (1979).
106. C. S. Paik Sung, T. W. Smith, C. B. Hu and Nak-Ho Sung, *Macromolecules*, **12**, 538 (1979).
107. R. Bill, M. Dröscher and G. Wegner, *Makromol. Chem.*, **179**, 2993 (1978).
108. Yu. L. Chichagova, A. A. Remizova, B. Ye. Myuller, N. V. Panova and N. P. Apukhtina, *Vysokomol. Soyed.*, **A17**, 2239 (1975) (in Russian). Translated in *Polym. Sci. USSR*, **17**, 2581 (1975).
109. G. A. Senich and W. J. MacKnight, *Adv. Chem. Series*, **176**, 97 (1979).
110. D. S. Huh and S. L. Copper, *Polym. Eng. Sci.*, **11**, 369 (1971).
111. J. Ferguson, D. J. Hourston, R. Meredith and D. Patsavoudis, *Eur. Polym. J.*, **8**, 369 (1972).
112. S. L. Cooper, J. S. West and R. W. Seymour, "Encyclopedia of Polymer Science and Technology", Supplementary Volume, Wiley, New York, NY, p. 521 (1976).
113. G. K. Hoeschele and W. K. Witsiepe, *Angew. Makromol. Chem.*, **29/30**, 267 (1973).
114. *Chem. Week*, p. 20 (July 6, 1977).
115. A. Ghaffar, I. Goodman and I. H. Hall, *Br. Polym. J.*, **5**, 315 (1973).
116. A. Ghaffar, I. Goodman and R. H. Peters, *Br. Polym. J.*, **10**, 115 (1978).
117. A. Ghaffar, I. Goodman, R. H. Peters and E. Segerman, *Br. Polym. J.*, **10**, 123 (1978).
118. H.-W. Hässlin, M. Dröscher and G. Wegner, *Makromol. Chem.*, **179**, 1373 (1978).

119. R. J. Cella, *J. Polym. Sci.: Polym. Symp.*, **42**, 727 (1973).
120. A. Lilaonitkul and S.L. Cooper, *Rubb. Chem. Technol.*, **50**, 1 (1977).
121. A. Lilaonitkul, J. C. West and S. L. Cooper, *J. Macromol. Sci.-Phys.*, **B12**, 563 (1976).
122. K. V. Van, S. L. Malhotra and L. P. Blanchard, *J. Macromol. Sci.-Chem.*, **A8**, 843 (1974).

CHAPTER 8

Other Block, Star and Graft Copolymers

Segmented molecules containing PTHF that have important commercial markets were discussed in Chapter 7. The polyurethanes were prepared by addition reactions of appropriate difunctional compounds. The polyesters were prepared by transesterification techniques. Even though the starting PTHF glycols are of nearly uniform molecular weight, the molecular weight distributions of the final materials are broad, especially compared to block copolymers prepared by anionic polymerization techniques.

Other mechanisms of polymerization have been used for the preparation of block, star and graft copolymers containing PTHF segments, and a wide variety of different materials have been made in the last 10 years. Efforts have been directed towards solving the experimental difficulties involved in their synthesis, and some clever methods of synthesis have evolved. Polymers with narrow molecular weight distributions can now be made. The systematic and detailed evaluation of properties that has been made with the commercial materials is lacking at present, and so it is too early to say whether any of the materials will have sufficient physical property and/or cost advantages over existing materials to lead to their commercialization. The synthesis and, when available, the physical properties of these other star and graft copolymers will be discussed in this chapter.

8.1 BLOCK COPOLYMERS

8.1.1 Block copolymers from caprolactam and poly(tetramethyl ether) glycol

Two slightly different techniques, method (a) and method (b), have been used to prepare block copolymers with segments derived from caprolactam and PTHF. The resulting polymers are quite similar except that one method leads to ABA terpolymers with ester end groups and the other to ABA terpolymers with lactam end groups. The starting materials for the syntheses are poly(tetramethyl ether) glycol (see Chapter 7 for a description of this material), caprolactam and terephthalic acid derivatives. The product has the general structure

"caprolactam block"—$[O(CH_2)_4—]_n O$—"caprolactam block",

where "caprolactam block" has the formula:

Endgroup 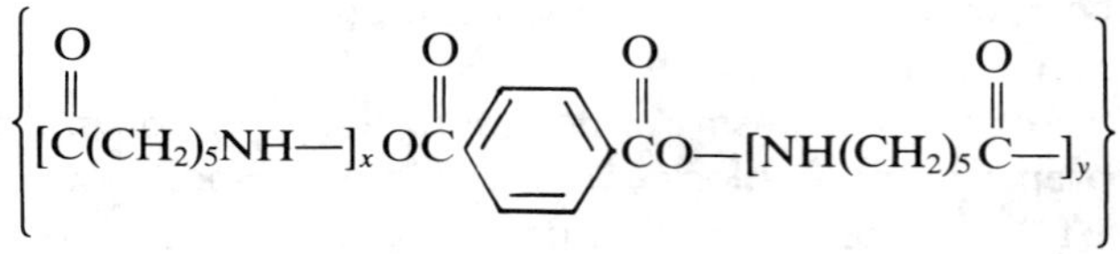

In method (a)[1,2] dry poly(tetramethyl ether) glycol is first treated with dimethylterephthalate or other carboxylic acid ester in the presence of a transesterification catalyst. Typical catalysts are mixtures of titanates and magnesium acetate. The prepolymer is a polyether-polyester of the structure

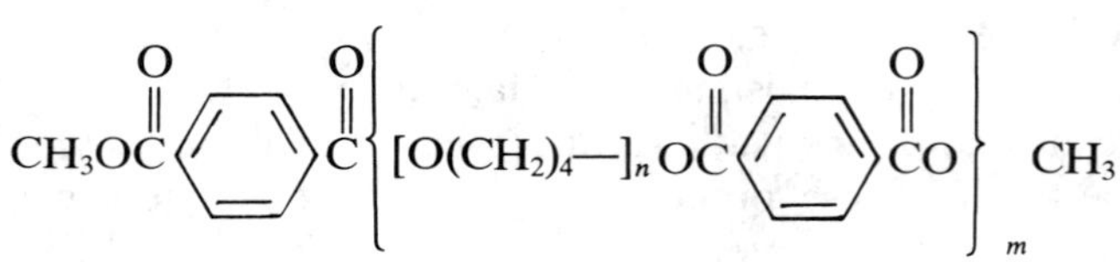

The anionic catalyzed polymerization of caprolactam, $\overline{NH-(CH_2)_5-C}=O$, with the polyester is initiated by adding an anionic polymerization catalyst like ethyl magnesium bromide. "Santowhite" powder, an antioxidant, is also added. In these polymerizations methyl ester end groups are formed and the polymer from caprolactam is inserted between the acid and alcohol moieties of the prepolymer.

Some typical tensile properties of terpolymers prepared in this way are given in Table 8.1. The changes with the molecular weight of poly(tetramethyl ether) glycol used and comparisons between terpolymers from poly(tetramethyl ether) glycol and other polyglycols are reminiscent of those given in Chapter 7 for polyesters containing PTHF. Further evaluation is needed to assess the utility of these terpolymers.

In method (b)[3] dry caprolactam is first converted to terephthaloyl-*bis*-caprolactam,

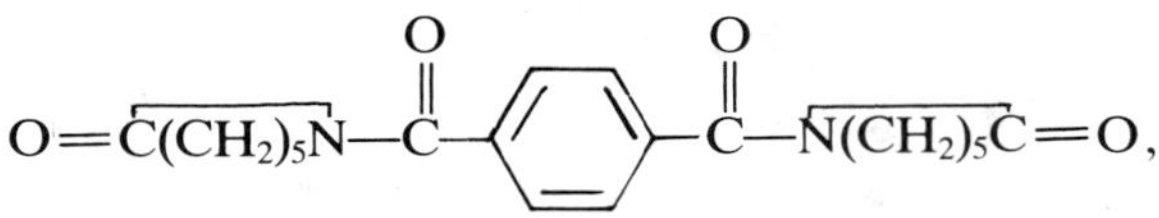

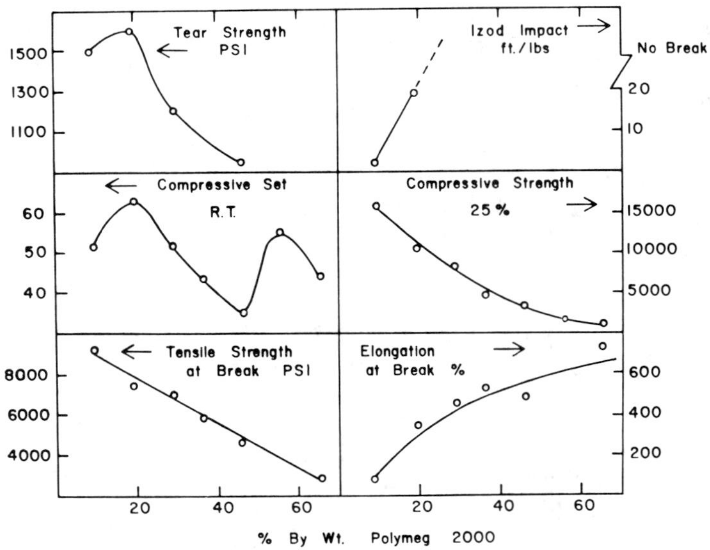

FIGURE 8.1 Properties of terpolymers from caprolactam and poly(tetramethylether) glycols prepared by method (b). The data were taken from Hedrick and Gabbert.[3]

TABLE 8.1
Typical Properties of Terpolymers from Caprolactam and Poly(tetramethylether) glycol[1–3]

			Tensile yield		Tensile fail		Tensile
Glycol	Method	Glycol, %	Strength, MPa	Elongation, %	Strength, MPa	Elongation, %	modulus, MPa
"Polymeg 2000"[a]	A	30	32	50	47	533	617
"Polymeg 1000"[a]	A	30	32	60	41	533	571
		40			36	776	177
"Polymeg 650"[a]	A	30			40	685	264
		40			21	754	106
"Polymeg 1000"[a]							
+ butanediol	A	21			19	156	45
		6					
PPG[b]	A	30			33	340	571
PEG[c]	A	30			48	520	393

[a]See Chapter 7 for a description of the properties of this poly(tetramethyl ether) glycol.
[b]"Voranol 2000", a polypropylene glycol.
[c]"Carbowax 4000", a polyethylene glycol.

by treatment with terephthaloyl chloride. Then the poly(tetramethyl ether) glycol, caprolactam, "Santowhite" powder and the terephthaloyl-*bis*-caprolactam are mixed together, and polymerization of the caprolactam is initiated by addition of bromomagnesium pyrrolidinone. Polymers prepared by method (b) have the same triblock structure as those prepared by method (a), but now the polymers have lactam end groups, $O{=}\overline{C(CH_2)_5N}{-}$. The variation of some properties of terpolymers prepared by method (b) with the content (wt.%) of the poly(tetramethyl ether) glycol block are shown in Figure 8.1. From the limited data shown, optimum properties seem to be at 20–30 wt.% of the glycol segment. Terpolymers prepared by either method (a) or method (b) can be crosslinked if thermosets are required.

8.1.2 Block copolymers from poly(tetrahydrofuran) and other poly(heterocycles)

Since polymerizations of PTHF are "living", the most obvious route to preparation of block copolymers containing PTHF is to add another monomer whose polymerization is amenable to initiation by addition of a THF oxonium ion and to allow the second monomer to polymerize. This method has been used repeatedly since the late 1960s. However, as discussed in Chapter 3, the comonomers that can be used are, for all practical purposes, only other heterocycles.

Copolymers from 3,3-bis(chloromethyl)oxetane (BCMO) In 1970,[4] Saegusa *et al.* prepared block copolymers derived from THF and 3,3-*bis*(chloromethyl)oxetane (BCMO). They first initiated the cationic polymerization of THF by adding BF_3 and epichlorohydrin, and then initiated the polymerization of BCMO from the growing end of the "living" PTHF formed in the first polymerization. Two types of block copolymers were prepared, AB and A(AB)B:

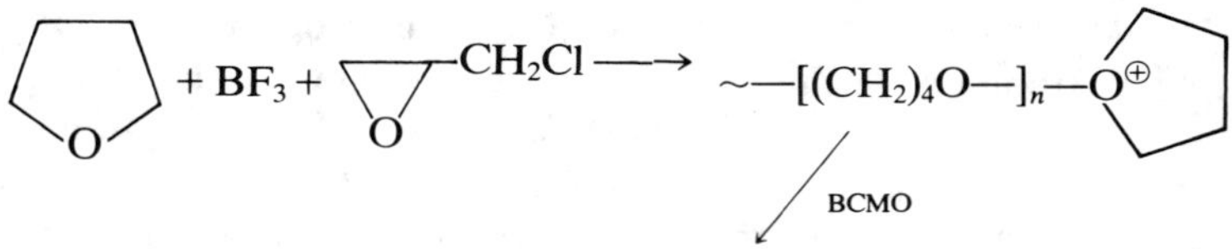

$$\sim\!—[(CH_2)_4O—]_{n+1}[CH_2C(CH_2Cl)_2CH_2O—]_m \sim,$$

$$AB\colon\ n+1=260,\ m=75$$

or

$$\sim\!—[(CH_2)_4O—]_{n+1}\{[(CH_2)_4O—]_y[CH_2C(CH_2Cl)_2CH_2O—]_x\}_z$$

$$[CH_2C(CH_2Cl)_2CH_2O—]_m \sim,$$

$$A(AB)B\colon\ n+1=240,\quad z=360\quad (x/y=46/54),\quad m=400.$$

One type of block copolymer was an AB type consisting of one PTHF (A) block and one poly(BCMO) (B) block. This copolymer was obtained only if, after addition of BCMO, the reaction mixture was rapidly cooled to −50°C, where polymerization is very slow, and excess of THF was removed under vacuum. When the reaction mixture was warmed to −30°C, polymerization of BCMO occurred.

If the excess of THF was not removed, the product was an A(AB)B block copolymer with crystalline blocks of PTHF (A) and of poly(BCMO) (B) bound to each end of a central rubbery block of THF–BCMO random copolymer (AB). Pure poly(BCMO) formed only after the supply of THF was exhausted either by copolymerization or vacuum stripping.[4]

Since these copolymers were prepared at 0°C and below, ceiling-temperature effects for THF polymerizations were insignificant. At higher temperatures it would be essential to cool the system before adding BCMO, if pure AB copolymer was desired. (For a discussion of the effect of comonomer on the THF monomer concentrations at equilibrium, see Chapter

3, where it was pointed out that THF can be consumed quantitatively in copolymerizations, even at temperatures where thermodynamic factors would limit its consumption in homopolymerizations.)

The lengths of the discrete blocks were calculated from the concentration of the propagating species determined by the "phenoxy end-capping" method (see Section 3.2.3). IR, DSC and solubility behavior confirmed the formation of the desired block copolymers. The AB block copolymer was pure, but the A(AB)B terpolymer was contaminated with some A(AB) block copolymer.[4]

Yamashita and Chiba[5] converted the hydroxyl groups of poly(tetramethyl ether) glycol to terminal dioxolenium perchlorate groups (see Chapter 7) and used the latter to initiate the polymerization of BCMO. An ABA block copolymer formed.

Copolymers from 7-oxabicyclo[2.2.1]heptane Initiation of polymerization of 7-oxabicyclo[2.2.1]heptane,

$$CH_2CH_2CHCH_2CH_2CHO,$$

by the dioxolenium initiator used by Yamashita and Chiba for BCMO gave block copolymer. However, poly(7-oxabicyclo[2.2.1]heptane) also formed as a result of transfer reactions.[5]

Copolymers from cyclic formals Attempted synthesis of block copolymers using the dioxolenium initiator above and dioxolane, $CH_2CH_2OCH_2O$, or tetroxane, $CH_2OCH_2OCH_2OCH_2O$, failed. Yokoyama *et al.*[6] found that cationic copolymerizations of 1,3-dioxolane with THF were sensitive to the counterion used. Block copolymers and/or mixtures of homopolymers formed when the initiator was triethyloxonium tetrafluoroborate.

Copolymers from pivalolactone Yamashita and Hane[7] prepared poly(tetramethyl ether) dicarboxylates from PTHF glycol by refluxing with a 70-molar excess of succinic anhydride for 48 hours. They used this product to initiate the polymerization of pivalolactone, $\overline{OCH_2C(CH_3)_2C}$=O, and form block copolymers.

Copolymers from ethylene oxide In 1972 Smith and Hubin[8] reported the synthesis of ABA block copolymers in which the B block was obtained from PTHF prepared using the difunctional initiator trifluoromethane sulfonic anhydride. The A blocks were obtained by adding ethylene oxide to the difunctional PTHF.

Block copolymers containing poly(oxymethylene) Chain transfer reactions have been suggested as a route by which block copolymers can form from pairs of poly(cyclic ethers).[9] Rosenberg *et al.* obtained either statistical or block copolymers by mixing PTHF and poly(oxymethylene) in benzene containing triethyloxonium antimonyhexachloride and heating at 50°C.

Properties of A(AB)B block copolymers The mechanical properties of the A(AB)B block copolymers obtained from THF and BCMO[4] were typical of a thermoplastic with crosslinks formed by crystallization of the PTHF block. Figure 8.2A shows a comparison of the room temperature stress–strain behavior of the block copolymer compared with "Thermolastic No. 125",[4] an ABA block copolymer based on poly(styrene), A, and poly(butadiene), B, and with vulcanized and unvulcanized SBR. Both block copolymers behave like vulcanized rubber. As is typical of thermoplastics with physical crosslinks, the strain at break, γ_b was a function of the temperature. Figure 8.2B shows that γ_b of the block copolymer falls quite sharply at about 40°C, the crystalline melting temperature of PTHF.

The A(AB)B block copolymer shows hysteresis behavior characteristic of a physically crosslinked block copolymer (see Chapter 7). Considerable hysteresis loss and a yield point occur in the first cycle of stretching but, as the cycle is

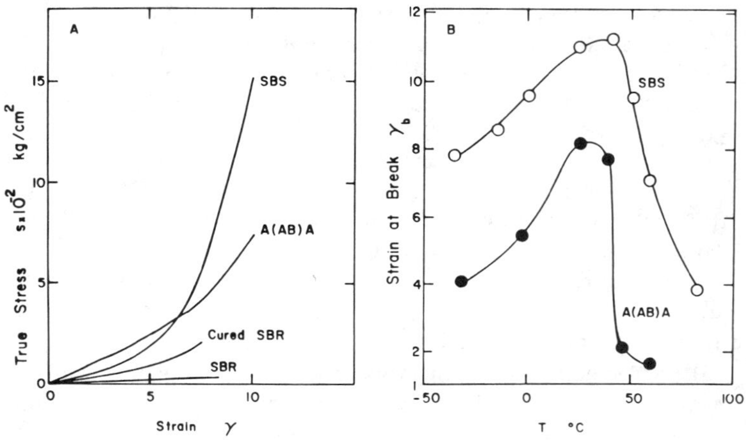

FIGURE 8.2 A: Room temperature stress–strain curves of A(AB)B block copolymers compared to other elastomers. The crosshead speed was 50 cm/min.[4] B: Strain at break, γ_b, of A(AB)B block copolymers as a function of temperature.[4] (Reprinted with permission from *Macromolecules*, **3**, 377 (1970). American Chemical Society.)

repeated, the hysteresis loss becomes quite small and the yield point disappears.[4]

8.1.3 Block copolymers from poly(styrene) and poly(tetrahydrofuran)

Very few block copolymers with a hydrocarbon block (derived from a vinyl compound such as styrene or from a diene such as butadiene) and a polyether block (derived from a polymerizable heterocyle such as tetrahydrofuran) have been prepared. Although the first of these copolymers was reported in 1966, the bulk of the development of synthetic methods has occurred since 1974. Not surprisingly, model systems based on poly(styrene) and poly(tetrahydrofuran), the best studied "living" anionic and cationic polymerizations, respectively, have been used almost exclusively. This survey will also feature these two polymers. When other polymers have also been tried in a given synthetic method, the results will be given briefly.

Successful syntheses of well-defined block copolymers from styrene and tetrahydrofuran must possess all of the following features:

1) the ability to prepare "living" poly(styryl) anions which are capable of continued growth at either one end (monofunctional) or both ends (bifunctional);
2) the ability to prepare "living" poly(tetrahydrofuryl) cations which are capable of continued growth at either one end (monofunctional) or both ends (bifunctional);
3) methods of combining the anions and cations by mutual termination or of initiating the polymerization of cationic species from an anionic block or vice versa.

As will be demonstrated below, all of the technology, with the exception of the initiation of anionic polymerization from a polyether backbone, is highly developed at present. One method of initiating anionic polymerization of a hydrocarbon from a polyether backbone has been reported and will be mentioned below but, since it has not been carefully studied, it is regarded as an exception here.

The usual symbols for block copolymers will be used. Thus AB consists of one block each of two different polymers A and B. ABA contains two blocks of A and one block of B in the sequence given and $(AB)_n$ contains alternating blocks of A and B. The lengths of A and B are defined but may vary. In addition, a monofunctional anion will be designated A^- and a difunctional anion will be $^-A^-$. Similarly, monofunctional and difunctional cations are assigned the symbols B^+ and $^+B^+$, respectively. In formulae for block copolymers, poly(styrene) is given the symbol S and poly(tetrahydrofuran) is T.

Synthesis of A^- For the purposes of the discussion below, only one reaction needs to be mentioned: initiation by an alkyl lithium such as *sec*-butyllithium[10,11]

$$s\text{-}C_4H_9Li + \underset{\displaystyle C_6H_5}{CH_2{=}CH} \longrightarrow s\text{-}C_4H_9{-}CH_2{-}\underset{\displaystyle C_6H_5}{CH^{\ominus}}Li^{\oplus} +$$

$$n\,\underset{\displaystyle C_6H_5}{CH_2{=}CH} \longrightarrow s\text{-}C_4H_9{-}\left[CH_2{-}\underset{\displaystyle C_6H_4}{CH}{-}\right]_n{-}CH_2\underset{\displaystyle C_6H_5}{CH^{\ominus}}Li^{\oplus}$$

Synthesis of $^-A^-$ Two syntheses have been used in the syntheses of the block copolymers:

a) *Initiation by sodium naphthalene*[12] In this case

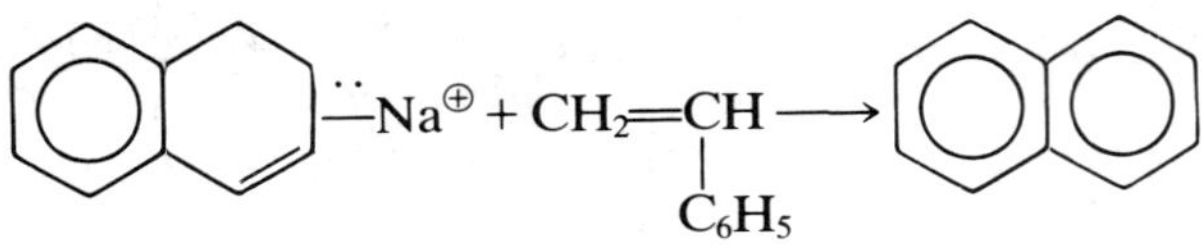

$$+ [C_6H_5\ddot{C}H\dot{C}H_2 \longleftrightarrow C_6H_5\dot{C}H\ddot{C}H_2]^{\ominus}Na^{\oplus} \longrightarrow$$

$$Na^{\oplus}\left[{}^{\ominus}\underset{\displaystyle C_6H_5}{CH}CH_2CH_2\underset{\displaystyle C_6H_5}{CH^{\ominus}}\right]Na^{+} + (n+m)\underset{\displaystyle C_6H_5}{CH_2{=}CH} \longrightarrow$$

$$Na^{\oplus}\left[{}^{\ominus}\underset{\displaystyle C_6H_5}{CH}CH_2{-}\left(\underset{\displaystyle C_6H_5}{CH}CH_2\right)_n{-}\left(CH_2\underset{\displaystyle C_6H_5}{CH}\right)_m{-}CH_2\underset{\displaystyle C_6H_5}{CH^{\ominus}}\right]Na^{\oplus}$$

b) *Initiation by α-methylstyrene tetramer*[10,11,13] The structure

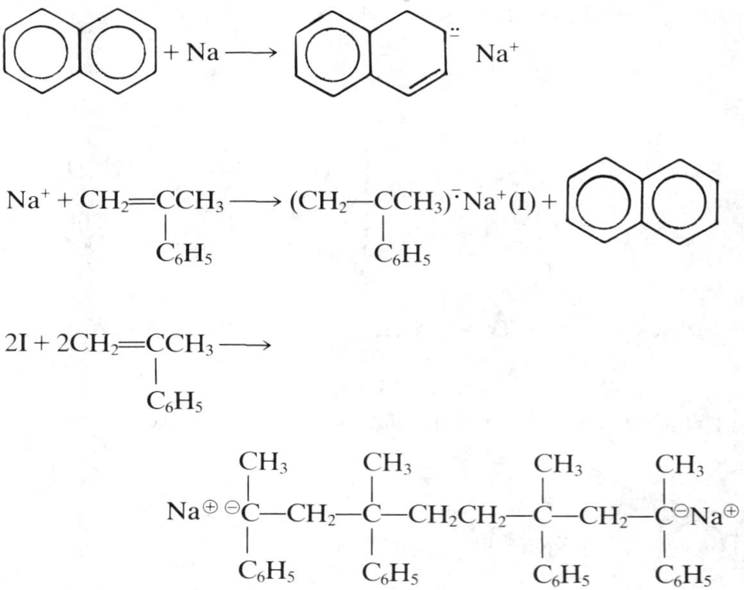

of the α-methylstyrene tetramer shown above was determined by Richards and Williams[11] and is the generally accepted one.

Synthesis of B^+ A wide variety of initiators will lead to a poly(tetrahydrofuran) that is capable of further polymerization on only one end (see Chapter 3). In order to ensure freedom from termination reactions at the usual polymerization temperatures (−10°C to +25°C), the choice of counterion is restricted to certain complex ions such as PF_6^-, SbF_6^-, $SO_3CF_3^-$ and, at the lower temperatures, BF_4^-. Since "living" polymers can be prepared from tetrahydrofuran, it is reasonable to suppose that poly(tetrahydrofuran) with a very narrow molecular-weight distribution can be prepared as readily as from poly(styrene). However, as was pointed out in Chapter 3 in the discussion of transfer, two factors make this more difficult. One is related to polymerization kinetics; in particular, the rate of depropagation is much faster for poly(tetrahydrofuran) than it is for poly(styrene). Calculations show that, as a consequence, poly(tetrahydrofuran) will reach a Gaussian distribution of molecular weights in about 20 days[14] rather than in the 80–100 years that has been calculated for poly(styrene).[15] This is not too serious a limitation because, at reasonable initiator concentrations, the rate of propagation is sufficiently fast to reach high conversions in 8–24 hours.

The other factor is related to the chemical constitution of the polymer itself. Every fifth atom in the poly(tetrahydrofuran) backbone is an ether oxygen, which is capable of undergoing transfer reactions with the growing center. Low temperature and low degrees of conversion minimize this transfer reaction, and quite narrow distributions were obtained by Ofstead[16] during polymerizations at 0°C with conversions in the range 5–40%. More recently, Croucher and Wetton[17] and Richards and coworkers,[18] polymerizing at −10°C, have succeeded in preparing near monodisperse poly(tetrahydrofuran) (typically, $\bar{M}_w/\bar{M}_n = 1.08$) with molecular weights of a few thousand to a million. In some cases, initiation was carried out at 95°C for four minutes prior to quenching the reaction mixture to polymerization temperature.[17] Since 95°C is above the ceiling temperature, this ensures that initiation will be completed before propagation ensues. The most suitable initiators include trialkyl oxonium ions,[19] alkyl trifluorosulfonates,[20] acyl halides in conjunction with silver salts[21–24] and certain alkyl halides in conjunction with silver salts.[23–30] The reactions that occur in each case were given in Chapter 3. As stated there, a whole host of other initiators have been studied, but the resulting molecular-weight distributions are less well known, and they have not been used to prepare block copolymers.

Synthesis of $^{+}B^{+}$ Only a limited number of ways of preparing a poly(tetrahydrofuran) that is active on both ends have been reported. These have already been presented in Chapter 7 in the section on difunctional initiators, where the important reactions were given. The initiators include 2,2′-octamethylene-*bis*-1,3-dioxolenium perchlorate (hereafter called *bis*-dioxolenium salt),[13,31] the anhydrides of trifluoromethanesulfonic acid (CF_3SO_3H) and of fluorosulfonic acid (FSO_3H),[20] PF_5,[32] terephthaloyl and adipolyl antimonyhexafluoroantimonates,[23,24] mixtures of *p*-xylylenedibromide and silver perchlorate or silver hexafluorophosphate[18] and a silenium dication initiator[33] $ClO_4^{\ominus}\,{}^{\oplus}SiMe_2\text{–}(OSiMe_2)_5\text{–}OMe_2Si^{\oplus}ClO_4^{\ominus}$. The same factors govern molecular weight distributions in the synthesis of $^{+}B^{+}$ as in the synthesis of B^{+}. Polydispersities of 1.3 to 2.0 are most common, but lower polydispersities have been reported for polymerizations carried out at $-10°C$.[18]

Synthesis of ABA block copolymers Two different kinds of block copolymer, TST and STS, must be considered. Their syntheses will be considered separately.

TST block copolymers The earliest report of the synthesis of block copolymers from poly(styrene) and poly(tetrahydrofuran) was in 1966 by Berger *et al.*,[34] who used ion coupling

$$Na^{+-}S^{-}Na^{+} + 2T^{+}BF_4^{-} \longrightarrow TST + 2NaBF_4\,.$$

Specifically, they prepared $(thf)_{20}(sty)_{24}(thf)_{20}$ and polymers of higher molecular weight. The successful synthesis of the block copolymer was demonstrated by solvent extraction and fractional precipitation. A small amount of homopoly(tetrahydrofuran) was also isolated. The source of this homopolymer was most likely termination with the BF_4^{-} counterion during the homopolymerization.[14]

In 1977 Richards *et al.*[18] reported a re-examination of the reaction between difunctional "living" poly(styrene) and monofunctional "living" poly(tetrahydrofuran):

$$Na^{+-}S^{-}Na^{+} + 2T^{+}PF_6^{-} \longrightarrow TST + 2NaPF_6\,.$$

They showed by a carefully worked out GPC (gel permeation chromatography) scheme[35] that the reaction "results virtually quantitatively in coupling of the two polymers". The reaction was not equally successful with poly(α-methylstyrene), where only partial linking occurred, or with poly(styrene) possessing "living" terminal 1,1-diphenylethylene units, where no linking at all was detected.

As will be noted below in discussing $(AB)_n$ block copolymers, Yamashita *et al.*[13] succeeded in coupling a difunctional anionically prepared poly(α-methylstyrene) with "living" poly(tetrahydrofuran) by first terminating the anionic polymer with CO_2.

In 1977, Hallensleben[36] reported yet another method for the synthesis of TST block copolymers. His method is quite similar to that of Lehmann *et al.*[24] and of Franta *et al.*[23] for the preparation of AB block copolymers and of poly(tetrahydrofuran) star polymers. Hallensleben treated bifunctional poly(styrene) bearing carboxyl groups with $SOCl_2$ to form the diacid chloride and used the diacid chloride in conjunction with silver salts and tetrahydrofuran to prepare TST block copolymer:

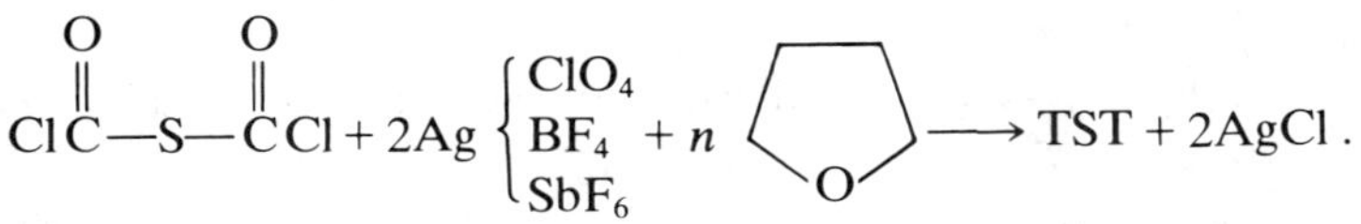

His only characterization consisted of an increase in number-average molecular weight. The same procedure was used to prepare ABA block copolymers from poly(α-methylstyrene).

STS block copolymers In 1971 Takahashi and Yamashita[37] reported the synthesis of block copolymers from difunctional PTHF and a poly(styrene) with a sodium carboxylate end group. The difunctional PTHF was obtained from the *bis*-dioxolenium salt and had perchlorate counterions:

$$ClO_4^{-\,+}T^{+}ClO_4^{-} + 2S\text{—}\overset{\overset{\displaystyle O}{\|}}{C}O^{-}Na^{+} \longrightarrow STS + 2NaClO_4\,.$$

Their polymers were contaminated with homopolymers, and

fractionation using THF–isopropyl alcohol mixed solvents was necessary.

Richards *et al.* have also re-examined the preparation of STS block copolymers by ion coupling.[18] For this purpose they used difunctional "living" PTHF prepared from *p*-xylylene dibromide and THF with the aid of $AgPF_6$ and treated it with a monofunctional poly(styrene) initiated by *n*-butyl lithium:

$$PF_6^{-\,+}T^+PF_6^- + 2S^-Li^+ \longrightarrow STS + 2LiPF_6\,.$$

The near quantitative efficiency of the coupling was again demonstrated by GPC as above. The GPC traces of each of the homopolymers and of the coupled product are shown in Figure 8.3. Both differential refractometer (DR) and ultraviolet (UV) detectors were used. The PTHF trace has a small UV signal resulting from the p-xylylene group, but the signal is very much smaller than the one due to polystyrene. The retention times of the homopolymers and of the coupled product are well separated from each other. It is evident that the coupled product, STS, has the expected molecular weight and that no detectable S^- remains after reaction with $^+T^+$.

A rather ingenious synthesis of STS block copolymer was reported in 1977 by Burgess *et al.*[29] They prepared "living" cationic ST^+ block copolymer by a procedure to be described below in the section on AB block copolymers and, by analogy to the sodium phenoxide end-capping procedure first described by Saegusa and Matsumoto,[38,39] coupled the block using disodium resorcinate:

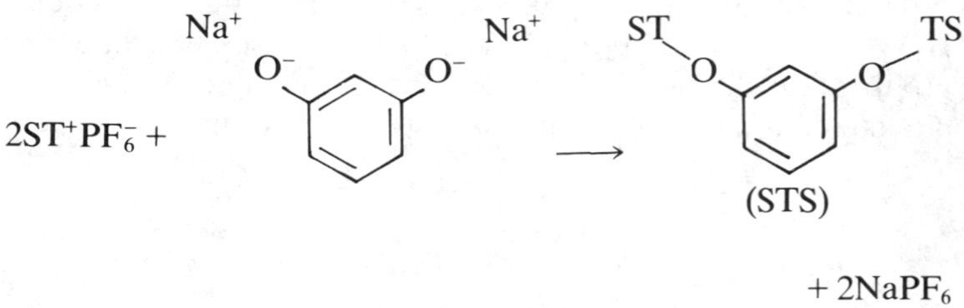

They demonstrated by GPC that the coupling efficiency was

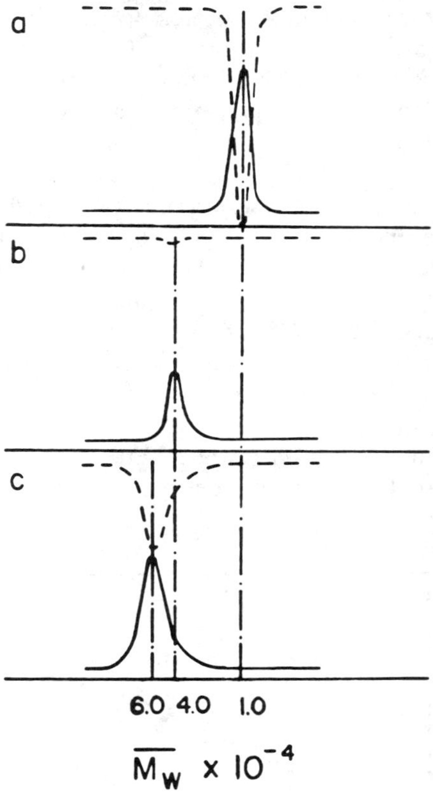

FIGURE 8.3 GPC traces of protonated S^- (a), methoxy-terminated $^+T^+$ (b) and STS (c), the product of reaction between S^- and $^+T^+$. The dashed curve corresponds to UV, 32X, and the continuous curve to DR, 8X.[18] (Reproduced from D. H. Richards, S. B. Kingston and J. Souel, *Polymer*, **19**, 68 (1978) by permission of the publishers, IPC Business Press, Ltd. ©)

about 70%. In addition, the block copolymer was contaminated by some homopoly(styrene) originating from the procedure used to prepare the ST block copolymer.

Properties of STS block copolymers Among the ABA block copolymers prepared by ionic methods of synthesis, physical properties other than those related to molecular weight and solubility have been reported for only STS block copolymers.[37,40,41] Yamashita and coworkers have examined morphology, rates of crystallization, surface tension, storage-loss moduli[37] and inverse gas chromatographic properties.[40] The number average molecular weight of the PTHF block was 72,000 and that of the poly(styrene) block ranged from 9,000 to 340,000. The block copolymers showed transitions characteristic of the homopolymers. Thus, as shown in Figure 8.4, which presents data obtained by dilatometry, the melting temperature of the PTHF block was nearly independent of composition and the T_g of the poly(styrene) block increased as the mole % styrene or, more significantly, the molecular weight of the poly(styrene) block increased. Dynamic mechanical measurements revealed that the T_g of the PTHF block was essentially independent of the mole% styrene of the

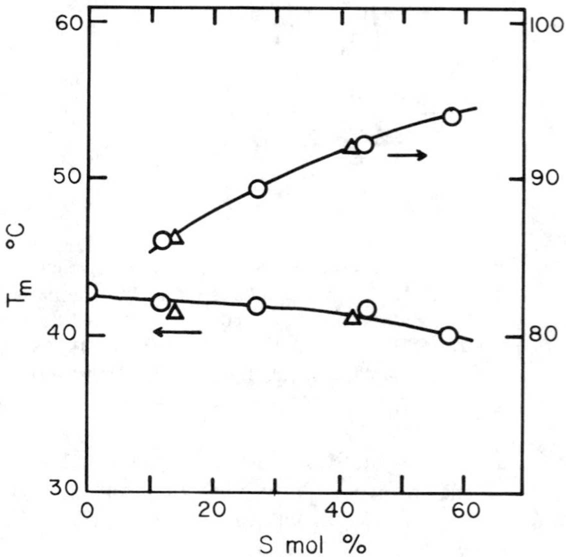

FIGURE 8.4 Transition temperatures of STS block copolymers as a function of mole % styrene, S.[37]

poly(styrene) block. The morphology of thin films cast from cyclohexane varied with composition. At high PTHF content, the films were spherulitic and lamellae were visible in electron-micrographs. The size of the spherulites decreased with decreasing PTHF content. At low PTHF content, the poly(styrene) was the continuous phase and the PTHF was present in spherical domains. Unfortunately, properties of STS copolymers with PTHF blocks of ~1,000 molecular weight were not studied. By analogy to the polyurethane and polyester commercial products the latter polymers would be expected to have the most useful properties.

The PTHF block had a surface activity which is 3 dyne/cm lower than that of the poly(styrene) block. PTHF tended to accumulate at the surface of the block copolymers even with low PTHF content in the block copolymers.[40,41] Thus the surface characteristics resembled those of pure PTHF.

Synthesis of AB block copolymers Most of the methods used above for the preparation of ABA block copolymers from poly(styrene) and poly(tetrahydrofuran) have also been applied to the synthesis of ST block copolymers. The purity of the products and methods of characterization were the same as described above, and will not be repeated here.

Takahashi and Yamashita[37] reported the preparation of ST block copolymers by ion coupling in 1975. They used a monofunctional polystyrene that had been treated with CO_2 and a poly(tetrahydrofuran) prepared using BF_3·THF complex.

Hallensleben[36] used a monofunctional poly(styrene)—or poly(α-methylstyrene)—carrying an acid chloride end group together with tetrahydrofuran and silver salts to prepare the ST block copolymers. Franta *et al.*[23] used a similar method but, instead of treating a poly(styrene) carrying carboxyl groups with $SOCl_2$, they prepared the acid chloride directly by terminating with phosgene and used that product along with $AgSbF_6$ and THF to prepare the ST block copolymer. They encountered two problems during the phosgenation. About 20% of the original poly(styrene) dimerized, and considerable termination occurred due to impurities in the phosgene.

Nevertheless, GPC showed that block copolymer had been formed.

Burgess *et al.*[28,29] used two different methods to add terminal bromide groups to monofunctional poly(styrene) prepared using *n*-butyl lithium. In both cases they then purified the intermediate and treated it with tetrahydrofuran and $AgClO_4$ or $AgPF_6$ to give the ST block copolymer. The first method of bromination involved termination with excess bromine:

$$S^{-}Li^{+} + Br_2 \longrightarrow S\text{–}Br + LiBr$$

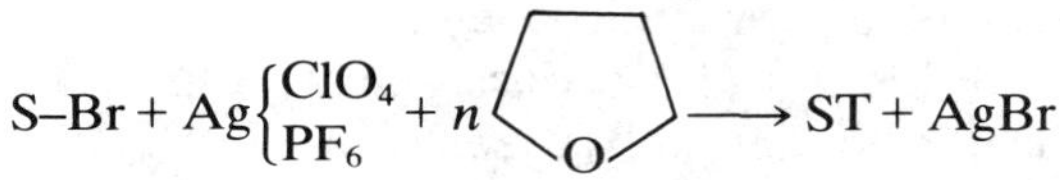

The transformation reaction was fair, but β-elimination to give an unsaturated end group was some problem. They found that using a polymeric Grignard reagent as an intermediate increased the yield of bromide to at least 95%. ST block copolymers with a polydispersity as low as 1.04 were prepared by this method. The poly(tetrahydrofuran) used for the narrow disperse polymer was prepared from $AgPF_6$, and tetrahydrofuran with initiation at −30°C for 30 min and propagation at −10°C for four hours. The use of ClO_4^- end groups and higher temperatures led to higher polydispersity.

The second method of bromination used excess xylylene dibromide:

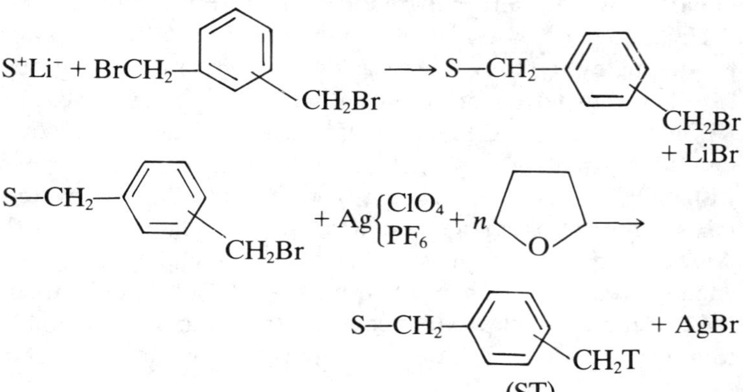

(ST)

The efficiency of the bromination was greatly improved by this method, but the resulting block copolymer had a higher polydispersity (1.1–1.17). Poly(butadiene-*b*-tetrahydrofuran) was also prepared by this method.

Richards *et al.*[18] used ion coupling. The monofunctional poly(styrene) was prepared using *n*-butyl lithium as initiator and the poly(tetrahydrofuran) was prepared at −10°C from benzyl bromide, $AgPF_6$, and tetrahydrofuran. Nearly quantitative coupling was observed.

Synthesis of $(AB)_n$ *block copolymers* All $(ST)_n$ block copolymers reported to date have been prepared by ion coupling of difunctional poly(styrene) or difunctional poly(α-methylstyrene) and difunctional poly(tetrahydrofuran). Yamashita *et al.*[31] used the *bis*-dioxolenium salt to prepare the poly(tetrahydrofuran) and an α-methylstyrene tetramer dianion to prepare the poly(styrene), which was terminated as above with CO_2. They prepared $(\text{thf}_{110}\text{–sty}_{53})_{30}$ in this way. They also prepared $(\text{thf}_{110}\text{–}\alpha\text{-methylstyrene}_4)_{34}$. In their experiments, attempts to couple the poly(styryl) anion directly or after addition of ethylene oxide gave either ketones or esters but no ion coupling. This could be an effect of counterion, since the byproduct salt that should form would be $NaClO_4$, which has been shown to ionize sufficiently in tetrahydrofuran to lead to polymerization in the presence of a reactive halide.[26]

Richards *et al.*[18] used ion coupling of difunctional poly(styrene) prepared from sodium naphthalene and styrene and difunctional poly(tetrahydrofuran) prepared from xylylene dibromide, $AgPF_6$ and tetrahydrofuran. They obtained nearly quantitative coupling. Note that the byproduct salt produced in their experiment is $NaPF_6$, which has been found to be inactive for tetrahydrofuran polymerization in the presence of reactive halides, probably because it does not dissociate.[26] This could provide the driving force for reaction here, whereas a similar reaction with different counterions failed above.

Kučera *et al.*[33] report the combination of a PTHF dication having perchlorate counterions with a poly(α-methylstyryl) dianion having sodium counterions. The dication was obtained using the selinium dicationic initiator and the dianion was the α-methylstyrene tetramer. The polydispersity coefficient was 2.9–3.5. According to these authors, the efficiency of combination is greater than 95% for low molecular weight ions (molecular weight about 1,000) and decreases with increasing molecular weight of the ions. Efficiencies of combination were always greater than 90%.

Note that an $(AB)_n$ block copolymer prepared from two difunctional polymers will necessarily have a rather broad (statistical) molecular weight distribution. This has been verified experimentally.[18,33] The GPC traces shown in Figure 8.5 show a very broad distribution for the (ST) copolymer. The ability to cross over at will from anionic to cationic polymerization and back again is needed to prepare $(AB)_n$ block copolymers of low polydispersity. This technology has not yet been developed. Methods of converting carbanions to growing oxonium ions exist, but methods of transforming the oxonium ions into carbanions still need to be demonstrated. Yamashita[41a] reports that the *bis*-dioxolenium salt is able to initiate polymerization of at least styrene as well as polymerization of cyclic ethers and cyclic formals, but there is as yet no report of the use of this initiator for the synthesis of $(AB)_n$ block copolymers. Because the initiation is known to be slow in the case of tetrahydrofuran polymerizations, it is unlikely that this initiator would be suitable for the preparation of $(AB)_n$ block copolymers of narrow polydispersity in any case. In addition, the mechanism of polymerization of the styrene is almost certainly cationic, and therefore complicated by transfer reactions to monomer that would eliminate successful coupling in a controllable manner.

Properties of $(ST)_n$ *block copolymers* Yamashita and coworkers have also studied the properties of $(ST)_n$ block copolymers.[37,40,41] The properties were very much like those reported by the same authors for STS block copolymers and described under that heading in this chapter. Some results are given in Figure 8.4. The molecular weight of the PTHF seg-

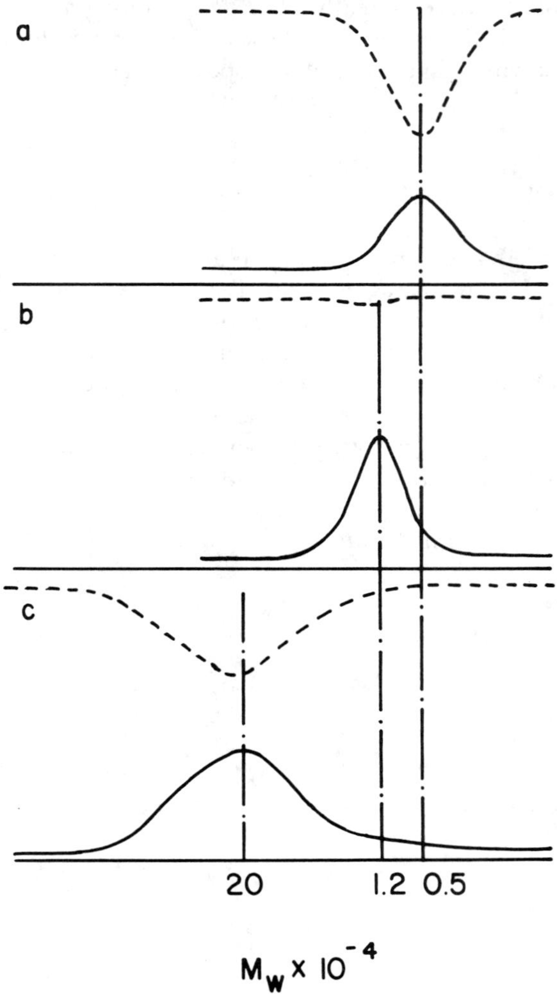

FIGURE 8.5 GPC traces of protonated $^{-}S^{-}$ (a), methoxy-terminated $^{+}T^{+}$ (b) and $(ST)_n$ (c), the product of reaction between $^{-}S^{-}$ and $^{+}T^{+}$. The dashed curve corresponds to UV, 32X, and the continuous curve to DR, 8X.[18] (Reproduced from D. H. Richards, S. B. Kingston and J. Souel, *Polymer*, **19**, 68 (1978) by permission of the publishers, IPC Business Press, Ltd. ©)

ment was not varied greatly. The same comments and conclusions reached for the STS block copolymers apply to $(ST)_n$ block copolymers and will not be repeated here.

8.2 STAR POLYMERS

Star polymers and copolymers from anionic polymers are well known and their characterization is well advanced. Counterparts in oxonium-ion polymerizations are nearly unknown. "One"- and "two"-arm stars, mono- and difunctional poly(tetrahydrofuran), respectively, have been reported frequently. The only report of the preparation of poly(tetrahydrofuran) stars with a greater number of arms was made by Lehmann *et al.*,[24] who used tri- and tetrafunctional aromatic acid chlorides with silver salts and tetrahydrofuran to prepare three- and four-armed stars.

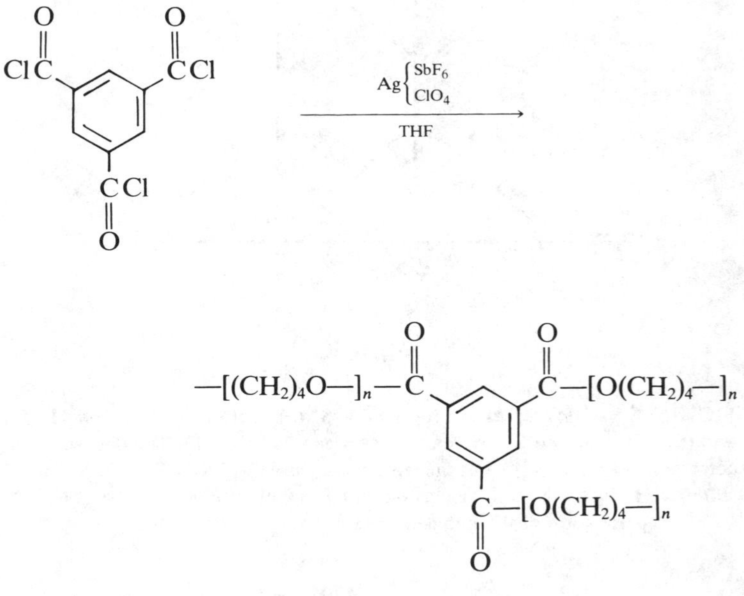

or

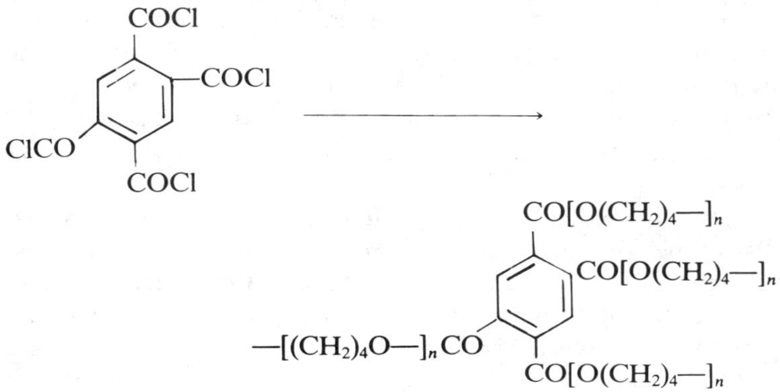

Initiation was efficient with the trifunctional acid chloride, and three-armed stars were readily prepared. Steric hindrance and nonuniform rates of initiation from the tetrafunctional acid chloride groups interfered with the preparation of four-armed stars. Kinetic data did not support successful synthesis, although uniform molecular weight branches were observed after hydrolysis of the ester linkages.

No attempts to prepare stars with block copolymer arms containing poly(tetrahydrofuran) are known as yet. However, a combination of the star technology already highly developed in anionic polymerization[42–50] plus ion coupling by some of the procedures above should make these polymers readily accessible.

8.3 GRAFT COPOLYMERS

PTHF constitutes the *branches* in all the graft copolymers containing PTHF which have been prepared by synthetic methods that permit some control over the number of branches.[23,26,51–57] Three methods of preparing such graft copolymers have been reported. In the first, a vinyl monomer with a side group capable of copolymerizing with THF is randomly copolymerized with another vinyl monomer and the polymerization of THF is initiated in the presence of the resulting copolymers. The polymerization may be initiated *from* the reactive group, copolymerization may proceed *through* the reactive group or polymerization may be terminated by reaction with the reactive group. When termination with the reactive group occurs, the term polymerization *onto* the backbone is applied.

In the second method, pendant allylic groups on the selected backbone, usually resulting from adventitious 1,2-copolymerization of a diene monomer but sometimes resulting from deliberate copolymerization with an appropriate comonomer, are converted to 1,2-epoxide groups. The graft copolymers are synthesized by copolymerization of THF *from*, *through* or *onto* these epoxide groups.

In the third method, allylic or other reactive halides either already present on the backbone or deliberately introduced serve as sites *from* which PTHF branches are grown by adding THF and a suitable salt. Suitable salts include many metal salts with complex counterions, especially silver salts,[23,26,61–66] and other salts such as $NOPF_6$.[67]

Examples of graft copolymers that have been prepared by each of these methods will be presented below.

Placement of the branches along the backbones is random for all three methods. The number of branches is determined by the statistics of the copolymerization reactions involved in the synthesis of the backbones. The number can be controlled to some extent by the amount of comonomer added or by reaction conditions. Generally, about 1 mole% of the total number of moles in a given backbone has reactive groups. In the third method, and sometimes in the other methods, when "living" PTHF chains are growing *from* the backbone, methods are available to determine the number of reactive sites in a given polymerization. Thus, graft copolymers with a known number of chains of known molecular weight randomly placed along the backbone can now be prepared. Regularly placed branches depend on the use of backbones with regularly placed active sites. Such backbones have not yet been used for the preparation of graft copolymers with PTHF branches.

PTHF constitutes the *backbone* in only one graft copolymer known to the author.[68] This graft copolymer was prepared by dissolving PTHF in vinyl chloride and then initiating the free radical polymerization of the vinyl chloride by addition of an appropriate initiator. Presumably a hydrogen atom is abstracted from the PTHF backbone leaving a free radical, which can combine with the free radical of the poly(vinyl chloride), grafting *onto* the PTHF, and/or can cause polymerization of the vinyl chloride *from* the PTHF. A statistical number and distribution of branches along the backbone results. The poly(vinyl chloride) branches are therefore of unknown molecular weight.

8.3.1 Graft copolymers prepared from copolymers containing reactive groups—the first method

Graft copolymers have been prepared from copolymers containing epoxy groups[51–54] by adding boron trifluoride etherate ($BF_3:Et_2O$). The backbones were vinyl copolymers of styrene, methyl methacrylate or acrylonitrile and glycidyl methacrylate ($CH_2{=}C(CH_3)CO_2CH_2\overline{CHCH_2O}$). Epoxidated phenol formaldehyde resins were also used as backbones. Grafting efficiencies, computed as (g of grafted PTHF/g of total THF converted to PTHF) × 100, were shown to be very high, 70–100%, but much or all of the graft copolymer was gelled. Such high grafting efficiencies indicate that in most cases very little homopoly(tetrahydrofuran) formed.

Yamashita and coworkers showed that the initiation of polymerization of THF by 2-phenyl-1,3-dioxolan-2-ylium perchlorate resulted in addition of the initiator.[55] Subsequently, in a multistep synthesis, Okada *et al.* prepared the corresponding polymeric initiator with 1,3-dioxolan-2-ylium ions attached to some of its phenyl groups and used this initiator to graft PTHF branches *from* the polystyrene.[56] The method is a cumbersome process and the growth of PTHF is accompanied by gel formation:

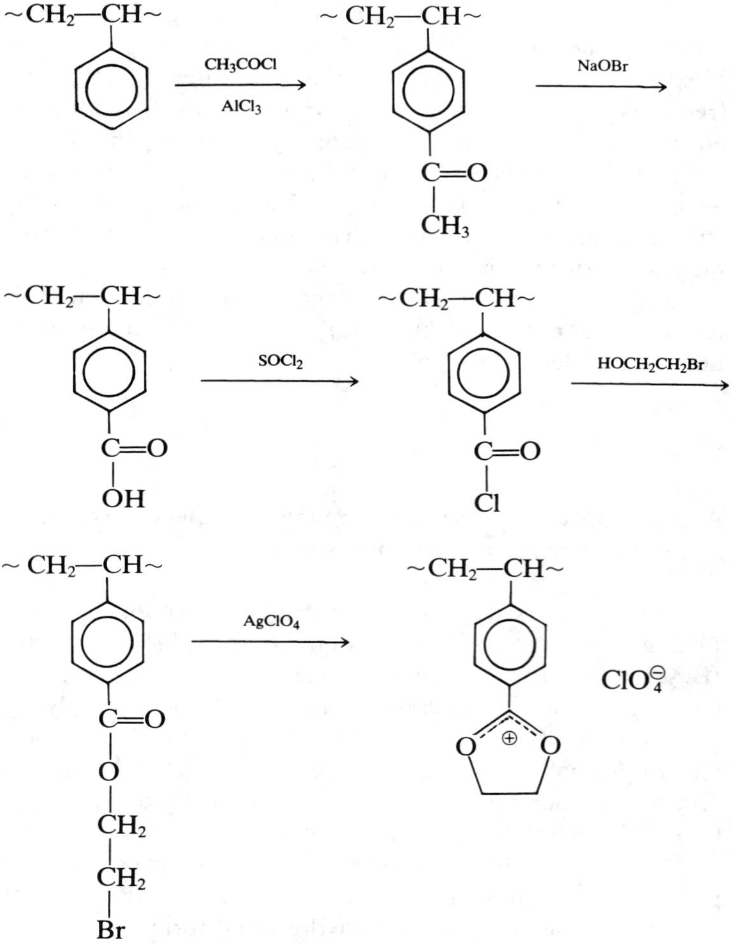

8.3.2 Graft copolymers prepared after epoxidation of pendant allylic groups on backbones

Epoxide groups on ethylene propylene terpolymers and epoxidized butyl rubbers have been used as sites for the addition of PTHF branches. The epoxide groups were introduced by reacting pendant 2-butenyl groups on Nordel 1440 or ethylidene norbornadiene groups in Epcar, both EPDMs (ethylene propylene terpolymers), or the copolymerized isoprene units in butyl rubbers with m-chloroperbenzoic acid. Graft copolymers were prepared by grafting *from*, *through* and *onto* the epoxide groups.[57–60] The initiator used was BF_3:THF or PF_5:THF in the grafting *from* method and $Et_3O^{\oplus}PF_6^{\ominus}$ in the other two methods. The copolymers were separated from unreacted backbone and homopolymer, if present, and were characterized. The analysis showed that, although grafting does occur by all three methods, the processes are complicated by side reactions such as self-condensation of the epoxide groups in the epoxidized rubbers. Depending on the structure of the epoxide, this self-condensation was either polymerization or dimerization. Only dimerization can occur with a trisubstituted epoxide of the structure obtained from copolymerized isoprenoid units. Model compound studies helped define the reactions occurring in each case. Grafting efficiencies varied from 50 to 100% depending on the backbone and method.

8.3.3 Graft copolymers prepared using polymers with reactive halide groups

This is the newest method and, of those proposed to date, the one most likely to lead to useful products. To prepare graft copolymers, the backbone is dissolved in THF and a suitable salt is added.

The backbones that have been used include poly(chloroprene), poly(vinyl chloride), chlorinated EPDM, chlorbutyl rubber, bromobutyl rubber, chlorinated or brominated poly(butadiene), chlorinated butadiene–styrene copolymer and polystyrene that has been copolymerized with chloromethylvinylbenzene or that has been treated in such a way as to introduce acyl groups.[21–24,26,62] Under ideal conditions, initiation is very efficient and the resulting graft copolymers are free from unreacted backbone and/or homo(polytetrahydrofuran). Thus, graft copolymers with a predictable number of branches of predictable molecular weight can be formed. In the graft copolymers prepared to date, the branches are randomly scattered along the backbone. Of the polymers listed, Neoprene W, a poly(chloroprene), is the preferred backbone in our group. Other backbones are more prone to side reactions, which result in low conversions of THF to polymer and/or gelled products even at low grafting efficiencies.

Polymerization is initiated by dissolving the backbone in THF and adding a suitable salt:

$$\text{Polymer–X}_n + \text{THF} + \text{MY} \longrightarrow \text{Polymer–(PTHF)}_n^{\oplus}\text{Y}_n^{\ominus} + \text{MX},$$

where X is a halide and MY is a metal salt with counterion Y. Studies have shown that suitable salt–halide combinations have the following characteristics:[26,29]

1) the salt is soluble in THF;

2) a stable oxonium ion–anion complex, $\sim\text{PTHF}^{\oplus}\text{Y}^{\ominus}$, is formed;

3) either the byproduct halide, MX, is insoluble, or at least its solubility is lower than that of the starting salt.

Halide reactivities are in the order: acyl > allyl ≥ benzyl > tertiary > secondary > primary ≥ 0.[26] Iodides are more reactive than bromides, which in turn are more reactive than chlorides.[30,70] Early data which suggested higher reactivity for chlorides, since graft copolymers were prepared more readily from chlorobutyl than from bromobutyl rubber, should be reinterpreted in the light of later results.[30] The cause of the poor grafting efficiency and PTHF yields with bromobutyl rubber was an elimination side reaction leading to dienes rather than to initiation of THF polymerization. Thus, it cannot be concluded *a priori* that greater reactivity of a given polymeric bromide necessarily means that the bromide is better than the corresponding chloride for the preparation of graft copolymers.

Suitable counterions, Y^-, include all the usual ones such as PF_6^-, SbF_6^-, BF_4^-, ClO_4^-, $CF_3SO_3^-$. All the usual comments summarized in Chapters 3 and 4 about stability versus temperature and so on apply.[14,71]

Among the metal salts, silver salts are preferred,[26,65,69] but mercury, lead, iron, aluminium and sodium perchlorates and lithium hexafluorophosphate can also be used. Silver salts are preferred for several reasons. In the absence of light or other activating reagent, PTHF polymerization does not occur. There are fewer side reactions related to the metal; high conversions to PTHF can be achieved. Pure graft copolymer can be prepared without extractions.

However, silver salts are not ideal. They are difficult to remove completely from the copolymers and then, as a result, the copolymers darken on standing. Some workers have resorted to techniques such as dialysis and ultracentrifigation to remove them. Others have washed with concentrated ammonia or sodium hydroxide solutions. Complexation with triphenylphosphine has been used. Ammonium hydroxide–ammonium chloride buffer solution with pH of nine seems to work better than ammonia alone. Besides darkening on exposure to light, the silver salts that remain in the copolymer can lead to insoluble copolymers, possibly as a result of metallation followed by undefined crosslinking reactions.

Among nonmetallic salts, nitrosyl hexafluorophosphate, $NOPF_6$, is the most promising.[67] The reaction that occurs is very similar to the one with metal salts, except that a covalent gaseous byproduct, NOCl, is produced instead of a metal salt:

$$\text{Polymer–X} + \text{THF} + \text{NOPF}_6 \longrightarrow \text{Polymer–PTHF}^{\oplus}\text{PF}_6^{\ominus} + \text{NOCl}\uparrow .$$

Polymers free from metallic residues and stable with time have been prepared with excellent efficiency using Neoprene W as the backbone.

Determining the number of branches in any graft copolymer is a major problem. This is true also for the graft copolymers prepared from reactive halide and THF. In theory, any of the methods described in Chapter 3 for the determination of the concentration of PTHF tertiary oxonium ions should be applicable. In practice, most of the methods are not sufficiently sensitive or cannot be used because functional groups in the reactants or in the backbone or the reactants themselves interfere. Eckstein *et al.*[72] compared the methods and determined the number of branches that form from Neoprene W. They found only two satisfactory methods: termination with triphenylphosphine followed by ^{31}P NMR and termination with NH_4OH–NH_4Cl buffer and reaction with fluorescamine followed by fluorescence spectroscopy. The fluorescence method was preferred. Both methods led to the conclusion that Neoprene W has 9 ± 1 active halogen atoms that are used to initiate the polymerization of THF when silver salts are used. Table 8.2 shows that, when this number is used, the agreement between calculated and experimental number average molecular weights of the graft copolymers prepared using $NOPF_6$ is very good.

The disadvantage of $NOPF_6$ is that, because of its high reactivity, reactant concentrations and reaction conditions must be rigorously controlled to avoid the formation of homopolymer from THF. Evidence for the formation of

TABLE 8.2
Number Average Molecular Weights of Polymers and Copolymers[72]

	Polymer 1	Polymer 2	Neoprene W
$[NOPF_6]$/[allylic chloride] (mol/mol)	17.4	19.4	—
THF/neoprene (g/g)	11	35	—
Pzn time (min)	8	15	—
PTHF in product, %	22	34	0
$\overline{M}_n$ [a]			
Experimental	212,000	243,000	180,000
Calculated	224,000	261,000	

[a]Based on the number of active sites determined by fluorescence spectroscopy.

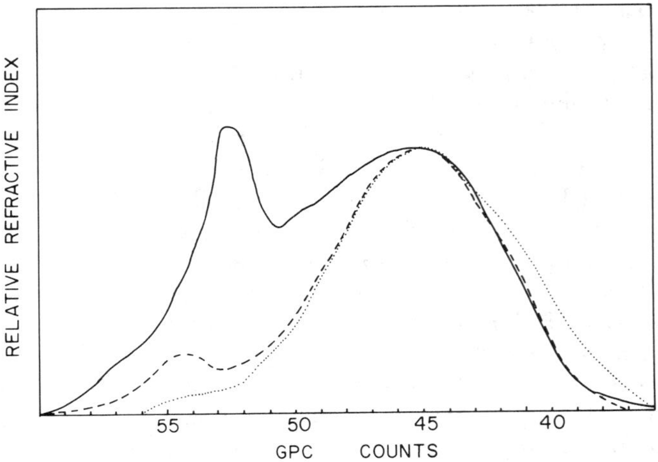

FIGURE 8.6 GPC of products of copolymerizations carried out in solution of THF/neoprene = 35 g/g and [$NOPF_6$]/[allylic chloride] = 19.4, terminated after 10 (dashed curve), 20 (continuous curve) and 180 min (dotted curve).[72]

homopolymer can be seen in the kinetics, DSC and in GPC traces such as those shown in Figure 8.6. After 10 minutes the copolymer peak is almost symmetrical, but after 20 and 180 minutes a second peak due to PTHF is clearly visible.

Graft copolymers have also been prepared using nitryl hexafluorosphosphate, NO_2PF_6, but the overall efficiency was poorer.[67]

8.3.4 Properties of Neoprene-*g*-poly(tetrahydrofuran) copolymers

Lehmann and Dreyfuss[63] reported the synthesis of Neoprene-*g*-poly(tetrahydrofuran) copolymers with branches of different length. The salt used in the preparation was $AgPF_6$. The composition of the copolymers ranged from 0 to 100% PTHF and the molecular weight of the branches varied from 7,600 to 150,000. Solubility, NMR and GPC indicated that pure graft copolymer was produced in all cases. Further characterization of the copolymer was later reported by Dreyfuss *et al.*, in a paper comparing the above graft copolymers with those of Neoprene-*g*-poly(7-oxabicyclo[2.2.1]heptane).[66]

The copolymers with PTHF branches behaved like homogeneous materials. They showed one set of transitions in both DSC and dynamic mechanical measurements. The variation of T_g and T_m is shown in Table 8.3. A 1:1 mixture of Neoprene W and PTHF showed two sets of transitions. Thin films cast from methylene chloride were spherulitic over the whole composition range when examined between crossed polaroids with the aid of an optical miscroscope. X-ray studies indicated that the PTHF crystallized over the composition range examined, 10–100% PTHF. This is consistent with results quoted in Chapter 7 for the polyesters. There it was reported that PTHF segments with molecular weights above 1,000 crystallized. In the graft copolymers being discussed here the lowest molecular weight of the PTHF branches was 7,600. Transmission electronmicrographs obtained from thin sections showed lamellae, as is seen in Figure 8.7.

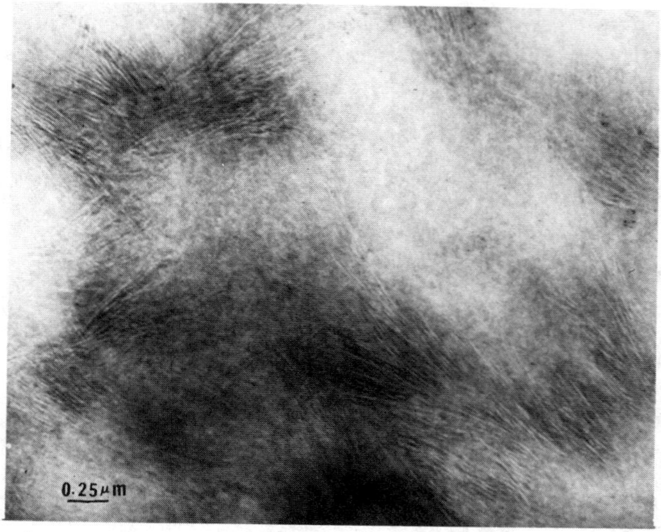

FIGURE 8.7 Transmission electronmicrograph of thin section of Neoprene W–*g*-PTHF containing 14% PTHF. The molecular weight of the branches was 7600.[66]

TABLE 8.3
Graft Copolymers of Poly(tetrahydrofuran) or Poly(7-oxabicyclo[2.2.1]heptane) from Neoprene[66]

Cyclic ether	δ, $(cal/cm^3)^{1/2}$	Branch, %	$\bar{M}_n$ of branch[a]	T_g [b]	T_m [b]
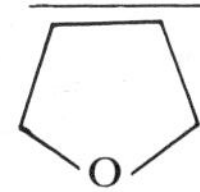	8.5–8.9[c]	100[d]	96000	−79	54
		90	150000	−71	53
		60	25000	−50	50
		34	7600	−46	51
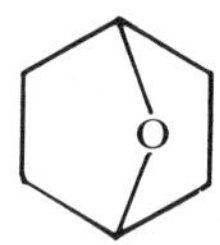	7.7[e]	100[d]	—	—	87, 175, 360
		59	24000	−43	40, 150, —
		13	2600	−37	46, 154, 335
—[f]	8.2–9.4[e]	0[f]	180000	−40	47
	—	Mixture[g]	—	−60, −15	28, 35

[a]Calculated from the number of branches per molecule determined by fluorescence spectroscopy and the amount of polymer of the cyclic ether determined gravimetrically or by ^1H NMR.

[b]In °C from DSC.

[c]Dreyfuss and Dreyfuss.[19]

[d]Homopolymer of the cyclic ether.

[e]Calculated using Small's molar attraction constants.[73]

[f]Neoprene W.

[g]A 1:1 mixture of poly(tetrahydrofuran) and Neoprene W cast from solution.

TABLE 8.4
Properties of Neoprene W–*g*-PTHF Copolymers[63]

PTHF, mole%	Strength at break[a] MPa	Elongation[a] at break, %	Permanent set after 16 hr, %
100	50	600	15
90	34	690	50
60	29	725	20
14	26	850	9.5
0	25	1150	5

[a]Crosshead speed was 5 cm/min.

TABLE 8.5
Adhesion of Neoprene AC Soft–*g*-PTHF Copolymers, Neoprene AC Soft and PTHF to Various Substrates[63]

		Peeling force on substrate listed, N/m [a]				
Sample	PTHF, %	PPO	Mylar	Aluminum	Chrome	Brass
Neoprene[b]	0	490	560	1220	170	480
14% branch[c]	14	370	760	840	220	290
69% branch[c]	69	310	780	850	300	280
PTHF	100	630[d]	400[d]	530[d]	120[d]	520[d]

[a]180° peel tests at 1 cm/min crosshead speed.
[b]AC Soft.
[c]See Table 8.3.
[d]All samples except chrome failed in a mixed cohesive–adhesive mode. Tests were run one to two days after preparation of the test specimens. Since PTHF may crystallize slowly even in the presence of the compounding diluents, there could be an important effect of aging.

Tensile tests of uncured specimens indicated that breaking elongation decreases and ultimate tensile strength increases as the length of the PTHF branches increases. The stress–strain curves were intermediate between those of the homopolymers. Some data are given in Table 8.4. Permanent set also increased with branch length and was thermally reversible. Adhesion to plastic and metal adherents in standard neoprene recipes was very similar to that of Neoprene AC, an adhesive-grade neoprene, or PTHF alone.[63] The data are quoted in Table 8.5.

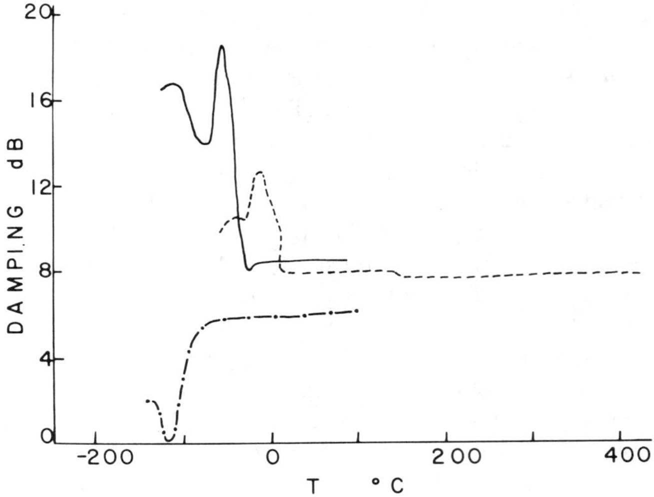

FIGURE 8.8 Dynamic mechanical analysis scans of damping against temperature for graft copolymers of polyethers from Neoprene W. The continuous, dashed and dash-dot curves correspond to copolymers containing 90% poly(tetrahydrofuran), 59% poly(7-oxabicyclo[2.2.1]heptane) and 34% poly(tetrahydrofuran), respectively.[66] Abbreviation: dB, decibel.

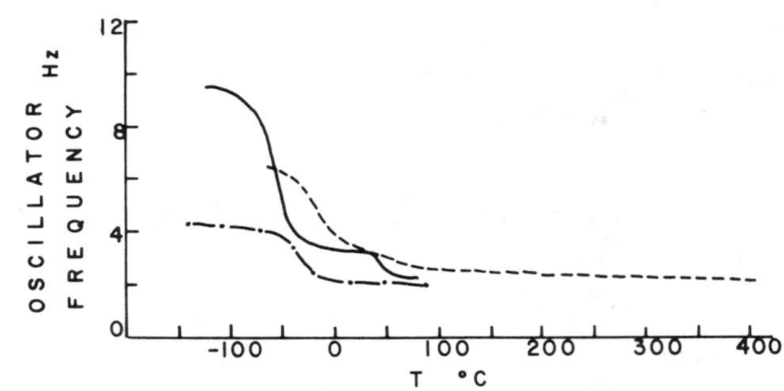

FIGURE 8.9 Corresponding dynamic mechanical analysis scans of oscillator frequency versus temperature for the graft copolymers in FIGURE 8.8.[66]

The graft copolymers with branches of poly(7-oxabicyclo[2.2.1]heptane) behaved more like the mixture of Neoprene W and PTHF. This is not surprising, since the solubility parameters for neoprene and PTHF are quite similar, but those for neoprene and poly(7-oxabicyclo[2.2.1]heptane) are far apart. The values are given in Table 8.3, which also shows that these copolymers had two sets of transitions, one for the branches and one for the backbone, although the transitions were shifted compared to those of the homopolymers.[66]

Figures 8.8 and 8.9 compare the dynamic mechanical analysis scans as a function of temperature for some of the graft copolymers listed in Table 8.3. At high content of polyether these scans exhibited damping peaks characteristic of the branches, but the peaks were somewhat shifted to higher temperature. For example, copolymers with PTHF branches had a damping peak below −100°C and a second damping peak slightly above the glass transition temperature of the graft copolymer. By analogy to the interpretation given in Chapter 6 for PTHF, the lower damping peak is probably associated with the onset of micro-Brownian motion of the main chain and the higher damping peak with a change to the rubbery state. No change in damping was detected at the melting temperature, but the scan of oscillator frequency versus temperature did show a transition that could be associated with the melting temperature. At low concentrations of the branches, the samples behaved mechanically like Neoprene except that the transitions were shifted to intermediate temperatures. This suggests an inversion of the continuous phase at an intermediate composition.[66]

8.3.5 Graft copolymers from poly(tetrahydrofuran) and vinyl chloride

The copolymers have been reported only in the patent literature.[68] The PTHF for their preparation can be obtained by any of the methods described in Chapter 3. From a synthetic point of view any molecular weight of PTHF can be used. Any composition can be prepared, but usually the polymerization of vinyl chloride was carried to high conversion. The PTHF was dissolved in monomeric vinyl chloride in the absence of air by stirring, usually at elevated temperatures such as 70°C for two hours. After cooling, the vinyl chloride was polymerized. Any conventional method including bulk, solution, suspension or emulsion could be used. For suspension polymerization, free radical catalysts such as *t*-butyl perpivalate or lauryl peroxide were added.

The method appears to be quite general. Copolymers of both THF and vinyl chloride were used. Sakomura *et al.*[68] made no attempt to fractionate the product into homopolymer and graft copolymer, and so the grafting efficiency is unknown. However, at least at concentrations of PTHF up to about 10 wt.%, the graft copolymers have good clarity. The clarity suggests high grafting efficiency, because blends of PTHF and PVC are not clear unless they are cast from solution and not heated.

Blends of PTHF and poly(vinyl chloride) or chlorinated poly(vinyl chloride) have been prepared by milling at 195°C on an oil-heated mill.[74–76] The high milling temperature was necessary to achieve good mixing. At 170°C, the highest temperature readily attainable on a steam-heated mill, the two homopolymers were essentially incompatible. Most probably, in the course of milling, sufficient graft copolymer similar to that described above was formed to compatibilize the homopolymers. Pure graft copolymer was not formed. Pressed sheets of these blends were opaque.

8.3.6 Properties of poly(tetrahydrofuran)–*g*–poly(vinyl chloride) and related polymers

Some physical properties of poly(tetrahydrofuran)–*g*-poly(vinyl chloride) copolymers are given in Table 8.6, which shows that 5–6% of PTHF or PTHF copolymer gave a PVC product with significantly improved impact strength. Compared to pure PVC, clarity was reduced, especially with the lower molecular weight PTHFs. Clarity was relatively better for the copolymer from vinyl chloride and vinyl acetate, but

TABLE 8.6

Properties[68] of Poly(tetrahydrofuran)–*g*-Poly(vinyl chloride) and Related Graft Copolymers[a]

PTHF,[b] $\bar{M}_n \times 10^{-4}$	PTHF,[b] wt.% in copolymer	Tensile impact[c] strength, kJ/m^2	Tensile strength,[c,d] MPa	Elongation,[c,d] %	Transmittance,[c,e] %
—	0	64	55	—	80
4	5.6	113	55	190	62
13.6	5.4	162	47	158	61
14.5	2.9	142	54	170	79
14.5	10.8	163	40	142	40
21	1.6	84	56	195	78
11.4	5.4[e]	118	46	—	76
36.0	5.7	127	46	—	70

[a]All the graft copolymers were prepared by dissolving PTHF in vinyl chloride monomer and then polymerizing the vinyl chloride in the normal way. Variable amounts of PTHF were used, but otherwise conditions were similar. A typical recipe included: vinyl chloride, 400 g; water (containing 0.2% deoxygenated hydroxypropylmethyl cellulose), 1,200 ml; *t*-butyl perpivalate, 0.56 g. Polymerizations were allowed to proceed at 60°C for 1.3 to 5 hours.

[b]All the PTHFs except that with $\bar{M}_n = 360{,}000$ were prepared using a THF : epichlorohydrin molar ratio of about 7 : 1 and initiation with triethyl aluminium and water. The polymer with $\bar{M}_n = 360{,}000$ was a homopolymer prepared using BF_3 as initiator.

[c]All physical properties were determined on 2 mm molded sheets from milled mixtures of: polymer, 100 parts; butyl-tin-mercaptide, three parts; butyl stearate, 0.5 parts.

[d]At break.

[e]At 650 mμ.

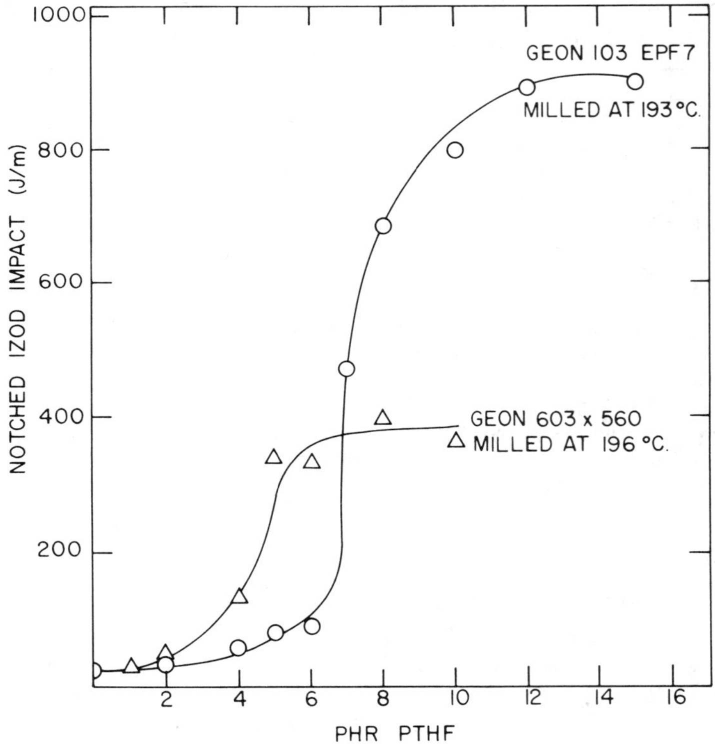

FIGURE 8.10 Effect of concentration of PTHF on notched Izod impact strength of PVC–PTHF and chlorinated-PVC(ClPVC)–PTHF blends. The recipes used were as follows.

Masterbatch for polyvinyl chloride

Material[a]	Parts by weight
Polyvinyl chloride (Geon 103 EPF7)	100.0
Processing aid[b]	3.0
Calcium stearate	2.0
TiO_2 (rutile)	5.0
Stabilizer[c]	3.0

(*caption continued on p. 290*)

Caption to Fig. 8.10 (*continued*)

Masterbatch for chlorinated polyvinyl chloride[a]

Material	Parts by weight
Chlorinated polyvinyl chloride (Geon 603 × 560)	100.0
Stabilizer[c]	2.5
TiO_2 (rutile)	5.0
Lubricant[d]	1.0

[a]Masterbatches were prepared by mixing in a Henschel mixer. Blends were then prepared by milling, on an *oil-heated* mill, the desired quantities of PTHF and masterbatch. Test specimens were compression molded.

[b]Styrene acrylonitrile copolymer.

[c]Dibutyltin thioglycollate.

[d]Low molecular weight poly(ethylene) "Ac 629A", made by Allied Chemical Corp. The intrinsic viscosity of the PTHF was 2 dl/g or greater (C_6H_6 at 25°C).[76]

impact strength and tensile strength decreased. The overall balance of properties was good.[68]

An earlier and more detailed study of the effect of the molecular weight and concentration of PTHF on properties of PVC was carried out on the blends described above.[74,76] Effects on elongation and tensile strength were similar to those given in Table 8.6. Tensile strength was somewhat reduced compared to pure PVC, and elongation increased. Thermal stability improved compared to other rubber modified PVCs, and impact strength was significantly enhanced. The major conclusions with regard to impact strength are summarized in Figures 8.10 and 8.11. As can be seen in Figure 8.10 for a PTHF of about 100,000 molecular weight (intrinsic viscosity of 3.2 dl/g in C_6H_6 at 25°C), high impact strength did not develop below about seven parts PTHF per hundred parts PVC (7 phr). Above that concentration, the impact strength did not change significantly, provided a high enough mixing temperature was used. Figure 8.11 shows that, at molecular weights below about 100,000, impact strength is not as easily developed even at proper mixing temperatures. Molecular

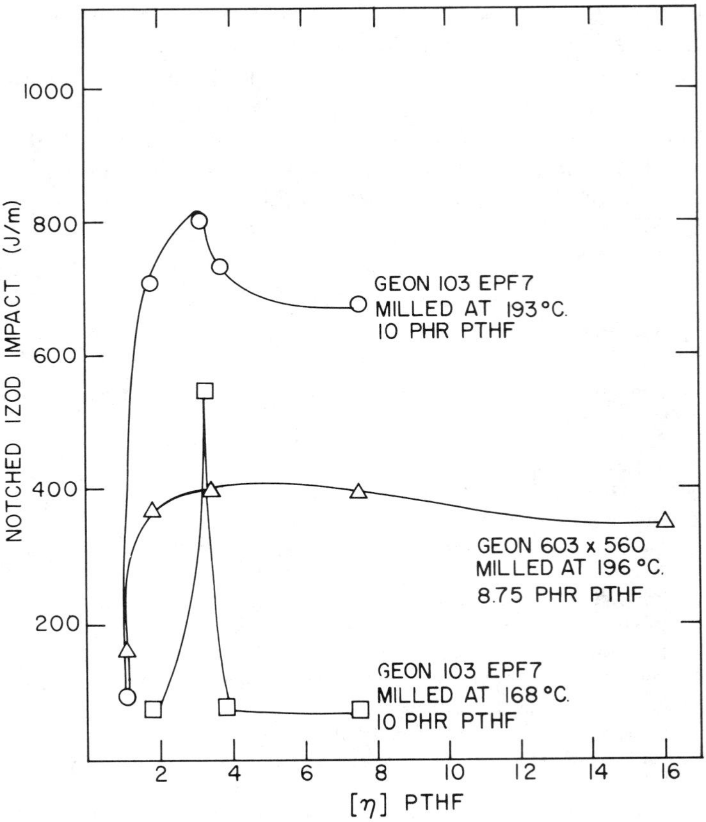

FIGURE 8.11 Effect of intrinsic viscosity of PTHF on the notched Izod impact strength of PVC–PTHF and ClPVC–PTHF blends. The recipes were the same as those used to obtain the data given in Figure 8.10.[76] The intrinsic viscosities, dl/g, were measured at 25°C in benzene.

weights above about 100,000 also improve impact strength, but the mixing temperature is more critical.

8.4 CONCLUSION

Although none of the block, star or graft copolymers described in this chapter is commercially important at the time of writing, some of these materials appear to have very desirable properties. Further evaluation, especially in areas such as abrasion resistance, needs to be made before a realistic assessment of their potentials can be made. The chemistry required to link PTHF to almost any type of material in a controlled manner is highly developed. One can anticipate that more of these materials will find commercial application in the future.

REFERENCES

1. R. M. Hedrick and J. D. Gabbert, US Patent 3,944,629 (to Monsanto Co.) (Mar. 16, 1976).
2. R. M. Hedrick and J. D. Gabbert, US Patent 3,993,709 (to Monsanto Co.) (Nov. 23, 1976).
3. R. M. Hedrick and J. D. Gabbert, US Patent 4,031,164 (to Monsanto Co.) (June 21, 1977).
4. T. Saegusa, S. Matsumoto and Y. Hashimoto, *Macromolecules*, **3**, 377 (1970).
5. Y. Yamashita and K. Chiba, *Polym. J.*, **4**, 200 (1973).
6. Y. Yokoyama, M. Okada and H. Sumitomo, *Polym. J.*, **11**, 629 (1979).
7. Y. Yamashita and T. Hane, *J. Poly. Sci.: Polym. Chem. Ed.*, **11**, 425 (1973).
8. S. Smith and A. J. Hubin, US Patent 3,644,567 (to Minnesota Mining and Manufacturing Co.) (Feb. 22, 1972).
9. B. A. Rosenberg, A. I. Yefremora and N. S. Enikolopian, *Vysokomol. Soyed.*, **7**, 2172 (1965) (in Russian). Translated in *Polym. Sci. USSR*, **7**, 2381 (1965).
10. S. Bywater, in "Comprehensive Chemical Kinetics", Vol. 15, C. H. Bamford and C. F. H. Tipper (eds.), Elsevier, Amsterdam-W., Netherlands, Chap. 1 (1976).
11. D. H. Richards and R. L. Williams, *J. Polym. Sci.*, **11**, 89 (1972).
12. M. Morton, *Encyclopedia of Polym. Sci. and Technol.*, **15**, 508 (1971).
13. Y. Yamashita, K. Nobutoki, Y. Nakamura and M. Hirota, *Macromolecules*, **4**, 548 (1971).
14. P. Dreyfuss and M. P. Dreyfuss, *Adv. Chem. Series*, **91**, 335 (1969).
15. A. Miyake and W. H. Stockmayer, *Makromol. Chem.*, **88**, 90 (1965).
16. E. A. Ofstead, *ACS Polym. Preprints*, **6**, 674 (1965).
17. T. G. Croucher and R. E. Wetton, *Polymer*, **17**, 205 (1976).
18. D. H. Richards, S. B. Kingston and T. Souel, *Polymer*, **19**, 68 (1978).
19. P. Dreyfuss and M. P. Dreyfuss, *Adv. Polym. Sci.*, **4**, 528 (1967).
20. S. Smith and A. J. Hubin, *J. Macromol. Sci.-Chem.*, **7**, 1399 (1973).
21. H. Meerwein, D. Delfs and H. Morshel, *Angew. Chem.*, **72**, 927 (1960).

22. G. A. Olah, S. J. Kuhn, W. S. Tolgyesi and E. B. Baker, *J. Am. Chem. Soc.*, **84**, 2733 (1962).
23. E. Franta, L. Reibel, J. Lehmann and S. Penczek, *J. Polym. Sci.: Polym. Symp.*, **56**, 139 (1976).
24. J. Lehmann, L. Reibel, E. Franta, W. Dobrogoszcz and S. Penczek, in "Abstracts to First International Symposium on Polymerization of Heterocycles (Ring-opening)", Warsaw-Jablonna, Poland, p. 141 (June 23–25, 1975).
25. F. Afshar-Taromi, M. Scheer, P. Rempp and E. Franta, *Makromol. Chem.*, **179**, 849 (1978).
26. P. Dreyfuss and J. P. Kennedy, *J. Polym. Sci.: Polym. Symp.*, **56**, 129 (1977).
27. F. J. Burgess, A. V. Cunliffe, D. H. Richards and D. C. Sherrington, *J. Polym. Sci.: Polym. Lett. Ed.*, **14**, 471 (1976).
28. F. J. Burgess, A. V. Cunliffe, J. R. MacCallum and D. H. Richards, *Polymer*, **18**, 719 (1977).
29. F. J. Burgess, A. V. Cunliffe, J. R. MacCallum and D. H. Richards, *Polymer*, **18**, 726 (1977).
30. K. I. Lee and P. Dreyfuss, *ACS Symposium Series*, **59**, 24 (1977).
31. Y. Yamashita, M. Hirota, K. Nobutoki, Y. Nakamura, A. Hirao, S. Kozawa, K. Chiba, H. Matsui and G. Hattori, *J. Polym. Sci.*, **B8**, 481 (1970).
32. R. Hoene and K.-H. W. Reichert, *Makromol. Chem.*, **177**, 3545 (1976).
33. M. Kučera, F. Bozek and K. Majerová, *Polymer*, **20**, 1013 (1979).
34. G. Berger, M. Levy and D. Vofsi, *Polym. Lett.*, **4**, 183 (1966).
35. F. J. Burgess, A. V. Cunliffe, J. V. Dawkins and D. H. Richards, *Polymer*, **18**, 733 (1977).
36. M. L. Hallensleben, *Makromol. Chem.*, **178**, 2125 (1977).
37. A. Takahashi and Y. Yamashita, *Adv. Chem. Series*, **142**, 267 (1975).
38. T. Saegusa and S. Matsumoto, *J. Polym. Sci.: A1*, **1**, 1559 (1968).
39. T. Saegusa and S. Matsumoto, *Polym. J.*, **1**, 31 (1970).
40. K. Ito, H. Sakakura and Y. Yamashita, *J. Polym. Sci.: Polym. Lett. Ed.*, **15**, 755 (1977).
41. Y. Yamashita, *J. Macromol. Sci.-Chem.*, **A13**, 401 (1979).
41a. Y. Yamashita, *Adv. Chem. Series*, **91**, 350 (1969).
42. J. Roovers and S. Bywater, *Macromolecules*, **5**, 384 (1972).
43. J. Roovers and S. Bywater, *Macromolecules*, **7**, 443 (1974).
44. N. Hadjichristidis and J. E. L. Roovers, *J. Polym. Sci.: Polym. Phys. Ed.*, **12**, 2521 (1974).
45. L.-K. Bi and L. J. Fetters, *Macromolecules*, **9**, 732 (1976).
46. N. Hadjichristidis, A. Guyot and L. J. Fetters, *Macromolecules*, **11**, 668 (1978).
47. R. N. Young and L. J. Fetters, *Macromolecules*, **11**, 899 (1978).
48. A. R. Luxton, A. Quig, M.-J. Delvaux and L. J. Fetters, *Polymer*, **19**, 1320 (1978).
49. B. J. Bauer and L. J. Fetters, *Rubber Revs.*, **51**, 406 (1978).
50. S. Bywater, *Adv. Polym. Sci.*, **30**, 89 (1979).
51. S. Aoki, T. Otsu and M. Imoto, *Kogyo Kagaku Zasshi*, **67**(6), 961 (1964); through *CA*, **62**, 6456 (1965).

52. T. Otsu, S. Aoki and M. Imoto, *J. Polym. Sci.*, **B2**, 223 (1964).
53. T. Otsu, M. Kondo, S. Aoki and M. Imoto, *J. Appl. Polym. Sci.*, **9**, 1991 (1965).
54. T. Otsu, M. Kondo, S. Aoki and M. Imoto, *International Chem. Eng.*, **6**, 711 (1966).
55. Y. Yamashita, M. Hirota, H. Matsui, A. Hirao and K. Nobutoki, *Polym. J.*, **2**, 43 (1971).
56. M. Okada, H. Sumitomo and T. Kakezawa, *Macromol. Chem.*, **162**, 285 (1972).
57. P. Dreyfuss and J. P. Kennedy, *J. Appl. Polym. Sci.: Appl. Polym. Symp.*, **30**, 153 (1977).
58. P. Dreyfuss and J. P. Kennedy, *J. Appl. Polym. Sci.: Appl. Polym. Symp.*, **30**, 165 (1977).
59. P. Dreyfuss and J. P. Kennedy, *J. Appl. Polym. Sci.: Appl. Polym. Symp.*, **30**, 179 (1977).
60. P. Dreyfuss and J. P. Kennedy, *J. Polym. Sci.: Polym. Symp.*, **60**, 47 (1977).
61. P. Dreyfuss and J. P. Kennedy, *J. Polym. Sci.: Polym. Lett. Ed.*, **14**, 135 (1976).
62. E. Franta, F. Afshar-Taromi and P. Rempp, *Makromol. Chem.*, **177**, 2191 (1976).
63. J. Lehmann and P. Dreyfuss, *Adv. Chem. Series*, **176**, 587 (1979).
64. E. Franta, L. Reibel and P. Rempp, *Polym. Preprints*, **20**(2), 102 (1979).
65. Y. Eckstein and P. Dreyfuss, *J. Polym. Sci.: Polym. Chem. Ed.*, **18**, 1799 (1980).
66. P. Dreyfuss, Y. Eckstein, D. P. Lee and J. Lehmann, *Polym. Preprints*, **20**(2), 94 (1979).
67. Y. Eckstein and P. Dreyfuss, *J. Polym. Sci.: Polym. Chem. Ed.*, **17**, 4115 (1979).
68. T. Sakomura, T. Yoshida, Y. Fujita and H. Shinbara, US Patent 3,696,143 (to Toyo Soda Manufacturing Co., Ltd.) (Oct. 3, 1972).
69. Y. Eckstein and P. Dreyfuss, *J. Inorg. Nucl. Chem.*, **43**, 23 (1981).
70. F. J. Burgess, A. V. Cunliffe and D. H. Richards, *Polymer*, **19**, 334 (1978).
71. P. Dreyfuss and M. P. Dreyfuss, in "Comprehensive Chemical Kinetics", Vol. 15, C. H. Bamford and C. F. H. Tipper (eds.), Elsevier, Amsterdam-W., Netherlands, Chap. 4, p. 259 (1976).
72. Y. Eckstein, D. P. Lee, R. P. Quirk and P. Dreyfuss, *J. Polym. Sci: Polym. Chem. Ed.*, **18**, 2021 (1980).
73. J. Brandrup and E. H. Immergut (eds.), "Polymer Handbook", Wiley, New York, NY (1966).
74. P. Dreyfuss and M. P. Dreyfuss, US Patent 3,548,037 (to B. F. Goodrich Co.) (Dec. 15, 1970).
75. P. Dreyfuss, US Patent 3,463,834 (to B. F. Goodrich Co.) (Aug. 26, 1969).
76. P. Dreyfuss, M. P. Dreyfuss and H. A. Tucker, *Adv. Chem. Series*, **128**, 125 (1973).

Subject Index

Author Index